THE
NEW TIME
TRAVELERS

THE

A JOURNEY TO THE

NEW TIME

FRONTIERS OF PHYSICS

TRAVELERS

DAVID TOOMEY

WITHDRAWN

W. W. NORTON & COMPANY

New York · London

Copyright © 2007 by David Toomey

For information about permission to reproduce selections from this book,
write to Permissions, W. W. Norton & Company, Inc.,
500 Fifth Avenue, New York, NY 10110

Manufacturing by Quebecor Fairfield
Book design by Charlotte Staub
Production manager: Julia Druskin

Library of Congress Cataloging-in-Publication Data

Toomey, David M.
The new time travelers : a journey to the frontiers of physics /
David Toomey.—1st ed.
p. cm.
Includes bibliographical references and index.
ISBN 978-0-393-06013-3 (hardcover)
1. Space and time. 2. Time travel. 3. Physics. I. Title.
QC173.59.S65T66 2007
530.11—dc22
2007011307

W. W. Norton & Company, Inc.
500 Fifth Avenue, New York, N.Y. 10110
www.wwnorton.com

W. W. Norton & Company Ltd.
Castle House, 75/76 Wells Street, W1T 3QT

1 2 3 4 5 6 7 8 9 0

Tell me a story.

In this century, and moment, of mania,
Tell me a story.

Make it a story of great distances, and starlight.

The name of the story will be Time,
But you must not pronounce its name.

Tell me a story of deep delight.

—ROBERT PENN WARREN
from *Audubon: A Vision*, 1969

CONTENTS

CONTENTS

CONTENTS

ACKNOWLEDGMENTS

everal of the figures who play important roles in this history took time from their work to share their recollections, and to offer perspective on events and ideas described in the pages following. Thanks are owed to David Deutsch, John Friedman, David Garfinkle, J. Richard Gott, Allen Everett, Thomas Roman, Kip Thorne, and Matt Visser; each is as gracious as he is generous. I am also especially grateful to Tom Roman for answering my many questions with care, commenting at length on several drafts of the manuscript, and offering valuable advice at various stages of the project.

I also wish to thank my colleagues in the English Department at the University of Massachusetts Amherst and the Humanities and Social Sciences Division at the University of the Virgin Islands. In particular, appreciation is owed to my friend and colleague John Nelson for encouragement and support throughout this project. Thanks to Larry Chunga, Stan Koehler, and Steve Peck for enlightening discussions of astronomy and cosmology; to Therese Hart and Mizan Kahn for answering questions of mathematics and its history; to Robert Kove for showing me many years ago that geometrical proofs may be "intuitively obvious"; and to Marla Miller for her suggestion of a title. For contributing photographs, arranging interviews, and locating archival materials, thanks to Tune Andersen, JoAnn Boyd, John Tepper Marlin, Christopher G. Oakley, Daniel Kuan Li Oi, and Marcia Tucker. Parts of this book were composed as the author was traveling; for opening their homes to me, I owe a debt of gratitude to Annette Markham and Elyzabeth Holford, and Patty O'Keefe and Scott Patterson. Much thanks to my students at the University of Massachusetts Amherst, whose intellectual curiosity is an ongoing inspiration, and who assure me that even in our practical-minded era there are still 2:00 AM residence hall discussions concerning the subjects of this book.

I am grateful to my editor, Angela von der Lippe, for taking a chance on this project and for telling me (in her inimitable and gracious way) when I was straying off course; to her assistant, Lydia Fitzpatrick; and to the capable staff of W. W. Norton for skillfully managing a hundred details of production. Insofar as the prose is clear and the presentation is accurate, much thanks to Stephanie Hiebert, who read, queried, and fixed. As always, warm thanks to my agent, David Hendin.

I am indebted to the staffs of the W. E. B. Du Bois and the Physical Sciences Libraries of the University of Massachusetts Amherst, the Young Science Library at Smith College, the Keefe Science Library at Amherst College, and the Historical Studies–Social Science Library at the Institute for Advanced Study. Finally, the books and articles whose titles appear in the bibliography of this book provided me a wealth of background material, and I refer the interested reader to them for more complete histories of theoretical physics and astrophysics in the last hundred years.

Of course, any errors are mine alone.

THE
NEW TIME
TRAVELERS

PROLOGUE

IN WHICH THE READER
IS REACQUAINTED WITH A CONVERSATION IMAGINED
IN A PREVIOUS CENTURY, AND THE AUTHOR'S
PURPOSE IS DESCRIBED

My first encounter with H. G. Wells's *The Time Machine* was innocent. When I pulled it from a library shelf, I had no idea it was regarded as a literary classic. I was drawn by the promise of that rather sensational title. When I opened to the first page of text and met with the line "The Time Traveller (for so it shall be convenient to speak of him) was expounding a recondite matter to us," there was no turning back. A few pages later Wells had painted a scene of late-Victorian-age gentlemen gathered in a drawing room, and I seemed to be among them.

In this scene the elegant and mysterious host is explaining the properties of the third dimension and theorizing about the nature of the fourth. There are excited discussions: a character identified as "the Psychologist" exclaims, "One might travel back and verify the accepted account of the battle of Hastings, for instance!" The "Very Young Man" says, "Then there is the future. . . . Just think! One might invest all one's money, leave it to accumulate at interest, and hurry on ahead!" A few minutes of polite debate, and suddenly the Time Traveller excuses himself. His guests hear the sound of his footsteps along the hall. Moments later he returns with an exquisitely fashioned model and carefully places it on a small table near a shaded lamp. His hand reaches toward the device, extends a long finger, and presses a miniature lever. There is movement of air, and a candle on the mantle flickers. The machine fades to an indistinct afterimage and holds for a moment. Then it vanishes completely. Save for the lamp, the table is empty. The Time Traveller informs his guests that the machine still exists. It has merely been sent to another time.

I finished the book in a few hours. Only years later did I learn that Wells intended the work as a social commentary on class division.[1] That it has come to be known as a science fiction primer on time travel is a deep irony. In my innocence, though, I was perhaps typical of the book's readership. My own interest lay not in the author's social theories, nor even with the strange people called Morlocks and Eloi. Rather, it was in the magnificent scenes near the end of the book, the last moments of a dying Earth, a dark creature crawling slowly on a beach, and on the sea's horizon, a swollen red Sun.

Still more intriguing was the possibility of time travel and the technology that might make it real. I read that opening conversation again and found more to appreciate. The skeptics were allowed a voice, and the Time Traveller anticipated and addressed their objections with grace and humor. His argument was reasonable, his demonstration persuasive. Although the machine's outline was left unclear, its details—a bit of ivory here, a bar of polished brass there—suggested an intricate mechanism. The Time Traveller drew his guests' attention to a bar that had about it "an odd twinkling appearance . . . as though it was in some ways unreal." We are never told what function the piece serves or why it seems already to half-inhabit another dimension, but the detail contributed to an effect of wonder. It was enough. A few more words were spoken, a finger depressed a tiny lever, and the machine vanished.

Of course, what was omitted in this leap from theory to demonstration was an explanation of *how* the model worked, exactly how it managed to enter and traverse a realm that the Time Traveller called the fourth dimension. Later in the book appear more detailed descriptions of a full-scale machine. The overall design belies the more carriagelike or sleighlike depictions of cinematic versions: it has a saddle and a sort of control panel with dials and levers, and the Time Traveller straddles it as one would a bicycle. Of the "feeling" of time travel, he leaves most to our imaginations: "I am afraid I cannot convey the peculiar sensations. . . . They are excessively unpleasant." But he explains that, much like riding a "switchback" (the British roller coaster), the experience gives a frightening sense that one is being hurled headlong through space, at every second about to crash.

Thus we are told what the machine looks like. We are told what it feels like to ride it. We are told all this. But we are never told exactly how a machine manages to travel through time. Of course, Wells did not know how. The conversation in the Time Traveller's drawing room was only a bit of narrative exposition to ground and make plausible the more fantastic episodes that would follow. I understood that. Nonetheless, my twelve-year-old imagination was slighted, and I felt the same dissatisfaction when, much later, I read the passage again. So, a few years ago I was fairly thrilled to learn that, in a manner of speaking, that drawing-room conversation had been continued, and continues still. Theoretical physicists and astrophysicists have given thought—rather serious and protracted thought at that—to the subject that Wells had glossed over: the *how* of traveling through time.

In the 1930s and 1940s, and then in the 1970s, one mathematician and three physicists made tentative inquiries into the subject of pastward time travel. The work of each was duly referenced and duly forgotten. The investigation began again in earnest in the late 1980s, but not before overcoming a formidable obstacle: doubts about its respectability. Scientists have professional concerns: dissertation and tenure committees to please, department chairs and college deans to impress, and reputations to tend. Moreover, their livelihoods are in many ways contingent on other people. In addition to holding lectures and writing papers, university scientists are expected to raise a substantial part of the funds necessary to support their research. The committees allocating those funds, in turn, are composed of people concerned with their own reputations; and a committee does not want to become known as naïve, spendthrift, or an easy mark. For all these reasons, physicists investigating the possibility of time travel in the 1960s and 1970s worried that such interests might be regarded as trivial. This is not to say that they resisted the subject entirely. In fact, they broached it with each other in conversation, and a few of them published papers that alluded to it, at least obliquely. But the conversations were informal, and the papers disguised their content (at least to a lay reader) with opaque phrases like "causality violation" and "closed timelike curves."

In 1988, much changed. In September of that year, *Physical Review Letters* published a three-page piece with a fairly sensational title: "Wormholes, Time Machines and the Weak Energy Condition." The paper made it clear that the subject was determinedly theoretical, and that no one was likely to construct a time machine anytime soon. To do so would require, among other things, enough energy to move masses the size of several suns at relativistic speeds, or an ability to pull submicroscopic wormholes from the quantum foam and inflate them by a factor of 10^{35} (that is, followed by 35 zeros)—and in either case, to create and manipulate a kind of matter that has never been observed directly. Nonetheless, the paper's publication marked a fairly significant moment. It was the first time anyone had proposed a means to undertake pastward time travel in the universe that we know. It was also the moment when the idea of time machines crossed from the realm of science fiction into the realm of science.

The paper sent a small shock wave through the community of

theoretical physicists. The subject of time machines had suddenly been made respectable. Almost immediately, what had been a trickle of publications grew into a steady stream, and time machines became the subject of panels at professional meetings. Those driving this inquiry were larger-than-life figures: among them were Frank Tipler, Stephen Hawking, and Caltech's Kip Thorne.[2]

In 1992 a workshop devoted largely to the subject was held at the Aspen Center for Physics in Aspen, Colorado. There, physicists were given the leisure to converse at length, and the freedom to consider the more philosophical aspects of time travel: the causal paradoxes that might result from changing history, questions of free will, alternate universes, and the nature of a civilization with the power to send signals pastward. In the years following, these ideas were more fully shaped and appeared in print. By that time the conversation had branched in many directions. Some papers explored the problems that time travel presented to logic; others discussed the challenges it presented to physics. Still others revisited ideas of the nature of spacetime, and bizarre concepts of branching pasts and futures. As this book goes to press, the most respected professional journals in physics have published well over two hundred papers on the subject of time machines, and the physicists themselves have produced ten or twelve related books.

The scientists described in these pages have expanded the conversation that Wells imagined, broadening it to include a variety of time machines and deepening it to call upon not merely geometry but relativity, quantum mechanics, and (at least tentatively) quantum gravity. They have also made the conversation more rigorous, proposing and testing hypotheses (not in laboratories but, in the manner of theoretical physicists, through "thought experiments"), then dismissing them outright or accepting them and building further hypotheses upon them.

Through all this, the ambience of that gaslit Victorian drawing room has survived. The Time Traveller's guests would not feel out of

place in the oak-paneled seminar rooms of Cambridge or Princeton, or those of Caltech's Bridge Laboratory. They might, though,
be somewhat disoriented at the other locales. Indeed, the physicists
who discuss the subject do so everywhere. They meet incidentally in
hallways outside classrooms, and more deliberately over coffee or
tea. They attend professional conferences where they sit on panels
and deliver presentations, and listen to presentations delivered by
others. In recent years they have come to telephone and e-mail
each other almost daily, discussing their own work and asking
about the work of others. They write, most often in collaboration.
They show early drafts of papers to colleagues, asking for suggestions and objections. Some post their work online before it appears
in print and revise in light of comments from colleagues who read it
there. Finally, they examine each other's published work with care,
and when they perceive an error they respond with a passion that
might surprise a nonphysicist.

Obviously there is an aspect of science fiction here. Time travel is
one of the largest science fiction subgenres, and the influence of
science fiction upon scientists is undeniable. Many who feature in
this book happily admit that they were first inspired to careers in
science through reading science fiction, but in due course most
found actual science far more exciting and satisfying. Many felt as
did Einstein, who once remarked that science fiction distorts science and gives people the illusion of understanding it without offering true understanding.[3] In accordance with such views, I had
originally intended to avoid science fiction in this book entirely, but
as work progressed I came upon more and more science fiction that
anticipated physicists' and philosophers' ideas of the nature of
time, illustrated those ideas in dramatic and colorful fashion, and
offered rich veins of thought and lucid explanations of the very
ideas that the physicists were tackling. It became clear that to ban-
ish science fiction from the narrative entirely would be foolish.

The physicists' influence upon science fiction was, as I expected, significant. To take a ready example, the wormhole—that hypothetical "shortcut through space"—was discovered by Ludwig Flamm as a solution to Einstein's field equations in 1916, and it was subsequently studied by Einstein and Nathan Rosen in the 1930s. It has since become indispensable to science fiction authors whose plots require reasonably fast interstellar travel. There are many such influences, and some of them are surprisingly direct. In 1974, physicist Frank Tipler published an idea for warping spacetime in the most prestigious journal of theoretical physics: *Physical Review D*. In 1979, science fiction writer Larry Niven appropriated the concept, as well as the article's rather unwieldy title—"Rotating Cylinders and the Possibility of Global Causality Violation"—for a science fiction story. In the spring of 1985, Carl Sagan was revising the manuscript for his science fiction novel *Contact* and wanted to be certain that his depiction of warped spacetime was accurate. He asked physicist Kip Thorne to read the manuscript. Thorne readily complied and suggested corrections that Sagan later incorporated into the novel.

Generally I had expected an influence of science on science fiction. What I had not expected, though, was a strong countercurrent— that is, a measurable influence of science fiction upon science. I offer two examples. After he answered Sagan's question, Thorne became increasingly intrigued by its implications, and his subsequent work on the problem led to the paper that set in motion the larger inquiry that is the subject of much of this book. A few years later, science fiction author and physicist Robert Forward worked up notes for a novel incorporating an idea that Thorne and colleagues had described in print, and he asked Thorne to read and comment upon those notes. Thorne did, and in the process he learned of one of Forward's ideas that he would cite in a subsequent article. As my research progressed, similar influences seemed to appear everywhere. I realized that if my intention was to tell the recent history of the idea of time travel, then to disregard the role of science fiction would make the narrative not only less interesting, but downright misleading.

Some doubts remained, however. We live at a moment in history when reality and fiction are often blurred, sometimes with potentially dangerous consequences. Recent surveys of Americans' knowledge of science suggest rather alarming lapses—these occurring at a time when threats to health and well-being seem to surround us. The list of such dangers is long and all too familiar. There are global concerns: climatic change and its consequences, chemical and bioterrorism, genetic engineering, the probable inadequacies of missile defense systems. There are more personal concerns: electronic surveillance, stem cell technology, mood-altering drugs.

It is true that misuses of science and technology are in some ways to blame for certain social and environmental ills. Yet it is also true that science and technology have never been more important for our long-term survival. A greater public appreciation for scientific inquiry would benefit us all. As Wells himself put it, "History becomes more and more a race between education and catastrophe."[4] With such matters in mind, I have worked to keep the distinction between fact and fiction as clear as possible. Accordingly, this narrative will allude to a work of science fiction only when a given scientist has acknowledged its influence on him, or when I believe it offers the best means to elucidate a certain theory or idea. So as not to interfere with the main narrative, I have tried to relegate the second sort of mention to the notes.[5]

In theoretical physics and astrophysics particularly, actual science may be considerably more difficult than science fiction. But engaging that difficulty is its own reward: if we are to appreciate warped spacetime, for instance, we must become reacquainted with geometry of the Euclidean and non-Euclidean varieties. Likewise, any discussion of time travel requires brief narrative excursions into astrophysics, special and general relativity, and quantum mechanics. This book will necessarily revisit those subjects. The road, then, will be difficult in some places. But we are likely to find greater satisfaction at its end.

Is time travel possible? Special and general relativity allow a kind of one-way time travel into the future. But what we usually mean by time travel—that is, travel into the past or "round-trip" travel into the future and back to the present—is quite another matter. As to the viability of *those* endeavors, the preponderance of opinion is determinedly agnostic. We simply do not know. Furthermore, most researchers agree that we *cannot* know until we achieve a better understanding of quantum gravity, and such knowledge may lie a decade or more in our future.

Does this mean that we must wait to think seriously about time travel? The figures in this book have not waited. As historians of science like to remind us, scientists are human and so possess human weaknesses. It is unreasonable to expect them always to work in a methodical, step-by-step fashion—proposing a theory to explain a phenomenon, devising tests for the theory, implementing the tests, and so on. We would ignore human nature by expecting them never to reorder the steps, never to get ahead of themselves. In fact, it may be argued that many advances occurred precisely *because* steps were taken out of sequence. The French mathematician Henri Poincaré remarked, "It is possible to contemplate the spectacle of the starry universe without wondering how it was formed: perhaps we ought to wait, and not look for a solution until we have patiently assembled the elements . . . but if we were so reasonable, if we were curious without impatience, it is probable we would never have created Science and we would always have been content with a trivial existence."[6]

Sometime in the near future, physicists will have a better understanding of quantum gravity. At that time perhaps there will appear another book about time travel and it will offer more definite answers to the questions raised here. It will show that certain lines of inquiry have proven viable, it will discard others, and it will introduce still others as yet unimagined. In the meantime, it is not too soon to recount what is known already, for that is a great deal.

The men and women described in these pages have thought seriously and deeply about a subject most of us left behind as children, in the process producing a discussion shot through with startling insights. In some ways, as they will admit, they may be getting ahead of themselves. But this is part of their appeal. Indeed, they would please Poincaré, for each, in some fashion, is unreasonable, curious, and impatient. Their story is the subject of this book, and its telling has three purposes: first, to trace the idea of time travel over several decades; second, to offer a view into the lives and work of a fascinating group of thinkers; and finally, to make long overdue payment on a debt owed a twelve-year-old boy—and, I hope, many others like him.

CHAPTER ONE

INTIMATIONS OF SPACETIME

IN WHICH WE MEET FANCIFUL BEINGS
WHO INHABIT THE SURFACE OF A PLANE, AND
THEIR REAL BUT NO LESS REMARKABLE INVENTOR

The rule is, jam tomorrow and jam yesterday, but never jam today.
— Lewis Carroll, *Through the Looking-Glass*

The Greek myths and the Old Testament have their prophets, and King Arthur's tutor Merlin could foretell events. Other myths and stories describe "frozen" time. Shakespeare's Hermione, for instance, exists in a kind of unaging suspension for sixteen years. Although actual travel into the past or future seems a more recent development of the human imagination, it does not lack for examples. At least twenty stories of time travel pre-date Wells's famous 1895 work. Among the better known are Charles Dickens's 1843 *A Christmas Carol* and Mark Twain's 1889 *A Connecticut Yankee in King Arthur's Court*. Dickens's Scrooge witnesses past and future in a dream and through the intervention of spirits, and the Connecticut Yankee's visit to medieval Britain is courtesy of a blow to the head by a falling crowbar. Less known pre-Wellsian works described time travel through suspended animation or hibernation, and enchantment by ghosts and fairies. All had something in common: in none of them could the time traveler control his travel—not his direction, not his rate of travel, not even when he began or ended that travel.[1] Late in the nineteenth century, though, two literary works described a means both to escape the present moment and to control that escape: clocks.

The second law of thermodynamics states that, over time, any closed system tends to become disorderly. Its energy is manifest in states that are less and less useful, and its atoms and molecules are distributed more and more randomly. A physicist would say that its *entropy* has increased. All closed physical systems are subject to entropy, and instruments that indicate the passage of time are no exception: candles cannot be made to "unmelt," and the sand in an hourglass cannot be made to trickle upward. A clock is, of course, subject to the same law: the potential energy of the wound spring is gradually transferred to the mechanical energy of the gears, and some is generated as heat energy through friction. Like all closed systems, a clock runs down. However, a clock may be reset; the hands on a clock may be wound backward. In short, a clock, or at least the dial of a clock, may be manipulated in such a way that makes it seem to defy entropy. For this reason a clock is especially useful to an author seeking to represent a means to control time, or to control one's travel through it.

According to literary historian Harry Geduld, the first appearance of a mechanism designed for time travel was in Edward Page Mitchell's 1881 short story "The Clock That Went Backwards," which tells of two boys who discover a clock that, when operated in the manner described by the title, allows them to visit the sixteenth century. Probably the second appearance of a time travel mechanism was in Lewis Carroll's 1889 *Sylvie and Bruno*, which featured a watch that not only indicated the time, but actually determined it. As the character called "the Professor" explains, "instead of *its* going with the *time*, time goes with *it*."[2]

Anyone objecting that these devices violate laws of nature would be something of a spoilsport, and such an objection would be rather beside the point. Like Dickens's spirits and Twain's crowbar, they are not presented as credible means of traveling through time. They are plot devices in the service of a fantasy with a time travel theme. Wells's *The Time Machine*, though, is a work of science fiction, a genre that by definition bases its plots upon known scientific phenomena or principles. We are justified in examining its science.

The Science of *The Time Machine*

The second half of the nineteenth century was witness to several new forms of travel: train, steamship, bicycle, and of course automobile, which made its London debut in 1895, the year of *The Time Machine*'s publication. If Wells's time machine seemed realistic to the novel's first readers, the reason was that it gave the appearance of being a logical next step in transportation. Of course, the machine itself is pure fantasy, and the particulars of its operation are quite properly left to our imagination. The principle behind that operation, however, offers somewhat more substance.

Wells's Time Traveller offers an idea of space that is part geometry and part metaphysics. The fourth dimension is temporal, he says, and is as real as the three dimensions of space—"length, breadth and thickness." Only because we enjoy great freedom of movement in the spatial dimensions, he says, do we recognize them at all. We do not conceive of the "time dimension" as real because we have no corresponding mobility within it. But the time dimension *is* real, the Time Traveller insists, and to persuade his guests of his conviction he engages them in a thought experiment. He asks whether an object that is instantaneous—that is, an object that endures for no time whatsoever—can be said to exist. Clearly, all agree, it cannot. To exist at all, an object must extend into three spatial dimensions, and also into a fourth, temporal, dimension.

By the waning years of the nineteenth century, intellectual circles were very much concerned with a *spatial* fourth dimension, and the idea had gained enough popular currency that charlatans were able to build a small industry around it.[3] Wells's Time Traveller though, takes the high road and legitimates his ideas by citing a Professor Simon Newcomb. Newcomb was no fiction, but a real contemporary of Wells—a professor of mathematics and astronomy at Johns Hopkins University who had published papers on higher dimensions.[4]

For these reasons it is interesting that, in his 1934 autobiography, Wells admits that the Time Traveller's speech was actually inspired by two rather different sources. At the Normal School of Science in South Kensington, a young Wells had participated in the students' Debating Society. There, he said, "I heard and laid hold of the idea of a four dimensional frame for a fresh apprehension of physical phenomena, which afterwards . . . gave me a frame for my first scientific fantasia, the *Time Machine*."[5] At about the same time, as editor of *Science Schools Journal*, Wells read a submission by a classmate who admitted that he had cribbed from an 1884 work, "What Is the Fourth Dimension?"[6] Its author was well known, and Wells's biographers assure us that Wells would have been familiar with his writings. Why, then, did Wells invoke Newcomb instead? Perhaps because, to name the real source, he would have been obliged to call to mind a scandal that would at the very least distract from the Time Traveller's story, and at worst discredit the very ideas he wanted his readers to consider seriously. In some ways it is a pity, as the man who provided Wells's actual inspiration was fascinating.

C. H. Hinton

He was a British mathematician named Charles Howard Hinton. Born in London in 1853, Hinton was a first son and showed mathematical promise at an early age. He matriculated at Oxford in 1871, and after two years he was offered a fellowship that enabled him to obtain the position of assistant master at the Cheltenham Ladies' College. As a young man, Hinton was much influenced by the progressive political and social ideals advocated by a group of intellectuals that included Havelock Ellis, George Boole (the inventor of Boolean algebra), and Boole's wife, Mary Everest Boole. Perhaps the most controversial member was Hinton's father, James, who urged the deliberate violation of social codes—including monogamy. In time, the younger Hinton married the Booles' daughter Mary, and with her he fathered four sons. But he seems either to have put his father's theories into practice or (as more than one wag put it) failed to grasp his father-in-law's concept of "either or." In 1886 the British courts learned that Hinton had sired twins with one Maud Florence Weldon, and on October 27 of that year he was put on trial for bigamy. Hinton's legal counsel expressed regret on his behalf, and the accused pleaded guilty, whereupon he was sentenced to three days in prison and released.[7] Then, evidently leaving Maud and the twins to their own devices, Hinton took Mary and his sons to Japan, where he gained employment at a middle school in Yokahama. His activities in the next few years seem lost to history, but by 1893 he turns up as an instructor of mathematics at Princeton University.

He taught at Princeton for four years before being fired, also for reasons unknown. But Hinton was a buoyant sort, and in subsequent years he held (however briefly) a variety of positions. Near the turn of the century he was hired by the U.S. Patent Office as an examiner of chemical patents. It was there that he remained until his death in 1907, which, as it happened, came without warning. Upon leaving the annual banquet of the Washington DC Society of Philanthropic Enquiry, Hinton collapsed of a cerebral hemorrhage and expired on the spot. His last public pronouncement, delivered at the banquet, had been a toast in praise of "female philosophers."

Hinton's intellectual legacy is diverse and rich. In the years between 1879 and 1907, he wrote extensively on various subjects and published ten books, several of which concerned other spatial dimensions. At the time Hinton was publishing, the subject was not new. In the early nineteenth century the German mathematician Carl Friedrich Gauss (about whom we shall learn more later) had considered the limitations of two-dimensional creatures he called "bookworms" or "flatworms" that lived on an infinitely thin sheet of paper.[8]

Many indulged in similar exercises. Perhaps the best known was a British schoolteacher named Edwin A. Abbott. His 1888 *Flatland: A Romance of Many Dimensions* is a fully satisfying tale with a plot that turns on—of all things—geometry.[9] Abbott's work is much imitated,[10] and its reputation has come to overshadow that of a work that Hinton published in the same year: a short piece called "A Plane World."[11] It described intelligent creatures confined to a two-dimensional surface, noted their limitations in perception, and concluded that we who inhabit a three-dimensional world might have analogous limitations. It further supposed that if the two-dimensional beings considered how a one-dimensional being might conceive of them, they might, by analogy, imagine our own three-dimensional world and we might, by the same analogy, imagine a world of four dimensions.[12]

Perhaps the most famous three-dimensional analogy of a four-dimensional form is the *tesseract* (Figure 1.1). The name, which may be Hinton's coinage, derives from the Greek *tesseres*, or "four."[13] As we will see, it is a lesson in our own limitations. In a one-dimensional universe, all that exists would move forward and backward along a single line. Everything in the universe would be restricted to that line; in fact, the universe would *be* that line. To enter a two-dimensional world or to create a two-dimensional form, we would simply move somewhere off that line. If we wished to create a square, we would project the line in a direction at right angles to itself, to a distance equivalent to its own length. To create a three-dimensional form—say, a cube—we would project a square at right angles to its edges for the length of an edge.

Logically, we would create a four-dimensional form by an analo-

Figure 1.1 A Tesseract

gous process—that is, by projecting the edges of a cube in a direction at right angles to themselves. But here we hit a cognitive barrier. We in the third dimension cannot imagine what direction that might be. What we *can* imagine, though, is a three-dimensional form that represents its four-dimensional counterpart much as a drawing of a cube represents an actual cube: that is, a tesseract. Not an actual four-dimensional form, but rather a sort of three-dimensional shadow of one, a tesseract appears as the framework of a cube with a second, smaller framework inside it. The smaller interior, appearing equally distant from any angle, is intended to represent an effect of perspective. If a tesseract were somehow transformed into the four-dimensional figure it represents, then the distance from any outer surface to that interior would be greater than the tesseract's diameter.

Hinton suggested that, just as a two-dimensional world may have a thickness—that is, a three-dimensional component undetectable to its inhabitants—so our three-dimensional space may have a slight four-dimensional thickness. In fact, he proposed that we could imagine such a space because some components of our nervous system are four-dimensional, and he further proposed that such imagining would have salutary effects. We may smile at this naïveté, but it is worth noting that on one particular Hinton was fairly prescient: "Our proportions [of the fourth dimension]," he said, "must be infinitely minute, or we should be conscious of them . . . it would probably be in the ultimate particles of matter, that we should discover the fourth dimension."[14]

The idea in some ways anticipates a feature of superstring theory: that dimensions higher than three are bound tightly at (very small) quantum scales. Hinton believed that much as three-dimensional space is embedded within four-dimensional space, so four-dimensional must be embedded within five-dimensional, five-dimensional must be embedded in six-dimensional, and so on.[15] On this score Hinton's reach may have finally exceeded his grasp. Although theoretical physicists working in superstring theory today are prepared to posit the reality of ten or eleven dimensions, they have found no reason to suspect an infinite regress.[16]

Hinton's ideas were prophetic in another detail. The science fiction trope of a wormhole—that is, a shortcut through spacetime—would be discovered by Ludwig Flamm as a solution to Einstein's field equations in 1916, and later studied by Einstein himself and his then student Nathan Rosen. Thirty years before Flamm published his work, the metaphor most commonly employed to depict a wormhole was used by Hinton, writing in 1886:

> Conceive of two beings at a great distance from one another on a plane surface. If the plane surface is bent so that they are brought close to one another, they would have no conception of their proximity, because to each the only possible movements would seem to be movements in the surface. The two beings might be conceived as so placed, by a proper bending of the plane, that they should be absolutely in juxtaposition, and yet to all reasoning faculties of either of them a great distance could be proved to intervene. The bending might be carried so far as to make one being suddenly appear in the plane by the side of the other. If these beings were ignorant of the existence of a third

dimension, this result would be as marvelous to them as it would be for a human being who was at a great distance—it might be at the other side of the world—to suddenly appear and really be by our side, and during the whole time not to have left the place in which he was.[17]

Hinton once commented on the interest of higher dimensions, writing, "It seems to me that the subject of higher space is becoming felt as serious. . . . It seems also that when we commence to feel the seriousness of any subject we partly lose our faculty of dealing with it."[18] No one would say that Hinton's own approach to such matters was completely serious—at least not always.

During his time at Princeton, Hinton invented a machine that would fire a baseball at forty to seventy miles per hour. One could regulate the ball's speed by moving an adjustable breech and determine its trajectory by turning rubber appendages inside the muzzle. Hinton first presented the machine before a gathering of the faculty and students, and no sooner had he begun speaking than he was interrupted by noises from the back of the room. There, a visitor who identified himself as a special-delivery postman was holding a letter addressed to the professor. Because Hinton had already been the victim of many practical jokes (he had been, in the term of the day, "horsed"), the audience expected that this was another such prank and prepared itself for a mild diversion. Hinton made a great show of protesting the interruption, but eventually he relented and accepted the letter, then begged the audience's indulgence as he took a moment to review it. He read aloud, and before he had gone more than a few paragraphs, the students realized that the joke was on them, and that its perpetrator was Hinton himself. He was reading a description of a baseball game played in the year 1950.[19]

It was, of course, a small joke about time travel. But Hinton did have serious ideas about the nature of time, and they were as unorthodox as his ideas about space. To appreciate them fully, we must examine some familiar experiences—so familiar that probably very few of us have thought much about them.

Experiencing Time

We put trust in our sense of space. Although we cannot see space itself, we can see objects that are arranged within it. We can move through its three dimensions at will, we can measure them directly, and we can even intuit dimensions and distances with some precision. Of time, though, we are far less certain. None of our five senses allow us direct contact with time, and intuition falters. You may know, for instance, that you have read ten pages since you began reading, and that you read, on average, a page every minute. By multiplying those numbers, you might guess that you have been reading for about ten minutes. But the measurement is indirect, and, as we all know, it may be wildly mistaken.

Sixteen hundred years ago, St. Augustine produced a cogent description of our difficulties in understanding time. "I know well enough what it is, provided that nobody asks me," he wrote, "but if I am asked what it is and try to explain, I am baffled."[20] Nonetheless he pressed further, saying that if nothing passed, there would be nothing he could call past time; if nothing were to happen, there would be nothing he could call future time; and if nothing were at all, there would be nothing he could call present time.[21] What Augustine described is formally called the "tensed theory of time." It divides time into past, future, and present. It is worth taking a moment to examine these divisions.

The Tensed Theory of Time

The tensed theory of time regards the *past* as made of events that no longer exist but have produced effects that may endure into the present and future. The past may have been influenced before it occurred or as it occurred, but it can be influenced no longer. In theory it is "knowable," but it may also be forgotten. The *future*, on the other hand, is made of events that have yet to come into being, of which we have meager knowledge or no knowledge, and over which we may exercise what is, at best, a limited control.

Although features of both past and future challenge under-standing, the *present* is a still greater mystery. It is subject to our moment-to-moment awareness; we are on most intimate terms with it, yet we barely know how to describe it.[22] It seems to have nothing that a physicist would call "internal structure"; that is, we can detect no moment or sequence of moments within it. Moreover, as the epigraph at the beginning of this chapter suggests, the pres-ent seems instantaneous: either it has no duration whatsoever, or its duration is too brief for us to discern subjectively or measure with instruments.

In a common, if unarticulated, understanding of time, the pres-ent is carried by time's flow and carries our conscious awareness, always in the same direction and always at the same speed. Events are like points along the shoreline that we approach, pass, and then leave behind us. With memories and records we can see some dis-tance behind, but we can only guess as to what is ahead. Contrarily, we might imagine that the present moment, with our conscious awareness, remains still, as though rooted in the riverbed. Events carried by the current pass *it*. This version seems particularly favored by writers. In *Metamorphoses*, Ovid observed, "Time itself, also, glides, in its continual motion, no differently than a river. For neither the river, nor the swift hour can stop: but as wave impels wave, and as the prior wave is chased by the coming wave, and chases the one before, so time flees equally, and, equally, follows, and is always new."[23] Marcus Aurelius had an unhappier take: "Time is a violent torrent; no sooner is a thing brought to sight than it is swept by and another takes its place, and this too will be swept away."[24]

Time and rivers have much in common—among them a hidden source and an end in a vast unknown. But beyond these rather superficial similarities, the comparison bears little scrutiny. It is of no help in answering the simplest of questions: how fast does time flow? That is, at what rate does a given moment move from the past, through the present, and into the future? Or, if we prefer, at what rate does a moment from the future move through the present into the past?[25] The obvious answer—one day every day, one hour each hour, one second per second—is unsatisfactory. If we define both "days" of "one day every day" as measures of the same time, then the answer is tautologous and meaningless, like my claim to have walked a kilometer for every kilometer I walked. If, on the other hand, we define the second "day" as a measure of a time that flows at a different rate than the first, and a time by which we are to measure the rate of the first, then we must ask at what rate the second time flows, and so must invent a third time by which to measure it, and so on. We have not solved the problem so much as we have displaced it, begging the original question into an infinite regress of other "times."

The Tenseless Theory of Time

In our day-to-day lives, most of us also operate with a "tenseless theory of time" in which all that has existed and will exist is located at some point along a continuum.[26] By its distance from a given moment, most commonly the year 0, any event may be located: Columbus's arrival in the Americas in 1492; the composition of this sentence at 10:30 PM on November 21, 2006; the closest approach of a small asteroid to Earth on April 13, 2029. This conception of time has no need for verbs, let alone verb tenses; indeed, it allows us to communicate our meaning clearly and accurately using neither.[27]

The tenseless theory deals with the problem of the present by simply dismissing it. It allows the present moment no privileged position other than that it is coincident with the moment at which one says, "This is the present moment," or thinks, "Now." But here our intuition objects. Surely the present moment *is* a privileged position. After all, we can see, hear, and otherwise sense things that exist in the present, whereas we cannot see, hear, or otherwise sense things that existed in the past or things that will exist in the future. Especially by such comparisons, the present seems *real*. To this, the proponent of tenseless time would answer that our perception is flawed. A moment in time is like a point in space. We cannot see, hear, or otherwise sense a moment in the past or future for the same reason that we cannot see, hear, or otherwise sense a certain street in Paris. We are not there.

The two theories are utterly incompatible, and it may be a tribute to the complexity of the human mind that most of us go about our lives using either or both as suits our purpose. With a mathematician's longing to simplify, however, Hinton envisioned a way to reconcile them. Imagine a sheet of still water. The air above the water surface represents the past, and the water below the surface represents the future. A stationary particle that endures through time appears as a stick passing at right angles through the surface. At the meeting of the film and the stick there is, of course, a membrane "collar" where the surface tension is broken by the stick, and where air ends and water begins. This point represents the present. As the stick is pushed downward through the membrane, it moves exactly as the stationary particle moves through the present moment. A *moving* object—say, a particle tracing a circle—would appear as a spiral.[28] The stick then represents the path of a particle through time; the spiral represents the path of a particle through time and space. Hinton called these paths "filamentary atoms."

Hinton's metaphor did not answer the problem of the duration of the present, nor of the speed at which it passes. It did, however, attempt a description of our conscious experience of the present moment. Hinton suggested that consciousness was connected with the membrane. This idea implied a limitation to our awareness. If a conscious person is moving in a circle, he is aware of his motion, but is unaware of the record of the successive intersections. He is unaware of the spiral that is his own filamentary atom.[29]

In a short story entitled "An Unfinished Communication," published in 1885, Hinton dramatized the idea. Its narrator, the unnamed main character, is wandering in an impoverished city neighborhood when he sees a signboard announcing, "Mr. Smith, Unlearner." He interprets it to mean that Mr. Smith relieves his clients of knowledge and its attendant burdens, and he realizes that it would be agreeable to forget, among other things, his formal schooling, which he calls (and here one suspects that the narrator is speaking for Hinton) "that plastering over the face of nature, that series of tricks and devices whereby [schoolmasters] teach a man knowing nothing of reality to talk of it as if he did."[30] So he seeks out the unlearner and soon, in a remote fishing village, finds him. As might be expected, the unlearner proves to be a rather strange man, and he describes an unusual notion of time. He claims that human consciousness perceives only a thin slice of existence called the present, although all the other moments continue to exist. Some weeks later the narrator is attempting to cross a shoreline at low tide. He finds himself engulfed in waves and, after much struggling, drowns. In the last moments of his life (and the first moments of his afterlife) he experiences episodes from his past immediately and most vividly, in a manner that demonstrates any past moment to be as real as any present moment, and shows that existence, when properly perceived, is an uninterrupted whole.

Hinton's ideas of space and time were not put forth as hypotheses to be tested, and they had no real grounding in physical law. To imply that they, in any meaningful way, anticipated the concepts that will be described in the next chapter would be to misread the history of physics, and no such suggestion is intended here.[31]

Nonetheless, it does seem that, in an approximate way, Hinton did predict some of those concepts, if only in outline. His filamentary atom foreshadows an idea that, some years hence, would be put forth by another university professor. And both his description of "folding" space and his suggestion that the present is illusory prefigure ideas that would be conceived by, as it happens, a rather more famous patent examiner.

CHAPTER TWO

EINSTEIN'S RADICAL IDEA

IN WHICH A FAMOUS SCIENTIST,
AFTER MUCH EFFORT, PROPOSES AN IDEA
FAR SUPERIOR TO EARLIER CONCEPTIONS; AND
SEVERAL MEANS TO TRAVEL TO THE FUTURE
ARE ILLUSTRATED

A picture without a frame is not a picture.
— John Archibald Wheeler[1]

I slam requires its adherents to pray five times a day facing in the direction of Mecca, a direction called the *qibla*. The *qibla* of any given point is a trigonometric function of its latitude, the latitude of Mecca, and the difference between the longitude of that point and the longitude of Mecca. Establishing the *qibla* for various locations took Islamic astronomers several centuries, and *qibla* tables are part of most Islamic astronomical treatises.

It appeals to a sense of symmetry to learn that, as the Islamic rules for prayer motivated the development of strategies to measure space, so the routines of Catholic monasteries—specifically the Benedictine monasteries of the twelfth and thirteenth centuries— forced the development of new ways to measure time. In winter months, matins (morning prayers) occurred several hours before dawn, and monasteries required that several monks keep vigil through the night, reciting a number of psalms in cadence to provide a reckoning of the time until that first office. In daylight hours the lack of a common time presented another problem. Monks working in the fields at some distance from the church had no easy way to know when the office of sext or none (pronounced like "known") was occurring. But the introduction of the clock meant that only one novice needed to be roused by an alarm so that he might wake the others; in daylight, bells could be rung to signal the seven periods of devotion, and monks would always know when to suspend their labors for prayer.

Many early clocks included elaborate astronomical representations, and the association of clocks with the heavens was much evident in late-medieval cosmology.[2] In the fourteenth century, Nicole Oresme, bishop of Lisieux, compared the universe to a vast mechanical clock created and set moving by God. In ensuing centuries the metaphor appeared again and again. The German astronomer Johannes Kepler used it, as did the British natural philosopher and theological writer Robert Boyle. It is easy to see why. A clock shared many characteristics with the heavens. It was precise; it moved smoothly, regularly, and predictably; and if it had a dial, its hands described circles. Moreover, because a properly maintained mechanical clock could tick for years without requiring resetting, it seemed to offer evidence that time existed independently of human timekeepers and human events.

Clockmakers' embellishments introduced new ideas of time and altered old ones. In 1336 a clock that featured a counting, striking train was installed in the church of San Gottardo in Milan. The train allowed it to count the hours—one bell at one o'clock, two at two o'clock, and so forth. A priest might be standing in the street before the church and a blacksmith might be at work on the outskirts of the city, and both could hear the sound of the same bells. It was a new experience—one that encouraged an idea of a single universal time, of a clock that was somewhere in space and whose time, or (let us be careful here) whose *broadcasting* of time was shared by all.

The 1687 publication of Isaac Newton's masterpiece, the *Philosophiae Naturalis Principia Mathematica* (commonly known as the *Principia*), codified this idea and grounded it in scientific language. "Absolute, true, and mathematical time of itself and from its own nature," Newton wrote, "flows equably without relation to anything external, and by another name is called duration."[3] In 1690, the philosopher and political theorist John Locke gave the idea an even clearer expression, writing, "This present moment is common to all things that are now in being, and equally comprehends that part of their existence as much as if they were all but one single being: and we may truly say, they all exist in the same moment of time."[4]

In our moment-to-moment lives, most of us would agree; we assume a single universal time and a single universal present. Asked to imagine a variety of places, we might picture the room that we are in as seen from our chair, the skyline of Manhattan as seen from New Jersey, Earth as seen from the surface of the Moon, or the Milky Way Galaxy as seen from somewhere in intergalactic space. If we are then asked to imagine all these views as they appear at present, we might place a ticking clock in one corner of each frame and make the time on all the clocks the same, or (perhaps) adjusted for time zones. In fact, this is the conception of time that many of us carry through our entire lives. It is, of course, utterly wrong. It would take a mind that was both brilliant and rebellious to intuit the far stranger truth.

Einstein

The childhood of Albert Einstein was rather typical of German Jews in the late nineteenth century. Born in 1879 to Hermann and Pauline Einstein, he attended public school (the Volksschule) and the Luitpold Gymnasium. He performed well; in fact, his grades in mathematics and Latin were often the highest in the class. Yet he chafed against the rigid program of learning and authoritarian instruction. His curiosity was better rewarded by a stimulating environment at home, where he enjoyed a harmonious relationship with his parents and his younger sister. Einstein's mother was a pianist, and from ages six to thirteen he took instruction on the violin. His uncle Jakob presented him with mathematical problems that he attacked with relish; and a family friend, a medical student named Max Talmud, joined the family for dinner every Thursday evening. Talmud brought the young Einstein books on many subjects; often they discussed philosophy and science.

On two occasions in childhood, Einstein had moments of understanding that were, in his own description, almost mystical. He wrote, "I experienced a miracle . . . as a child of four or five when my father showed me a compass." It so excited him that he "trembled and grew cold" and realized "there had to be something behind objects that lay deeply hidden."[5] When Einstein was twelve, he was given a textbook on Euclidean geometry, of which he would later write, "Here were assertions, as for example the intersection of the three altitudes of a triangle in one point, which—although by no means evident—could nevertheless be proved with such certainty that any doubt appeared to be out of the question. This lucidity and certainty made an indescribable impression on me." He came to call it the "holy geometry book."[6]

As a student at the Zurich Polytechnikum, Einstein developed a distaste for received scientific dogma that several of his teachers mistook for indifference. In some ways it was a difficult time. His parents were experiencing financial hardships, and he was deeply concerned for their welfare. But he had friends: fellow students Marcel Grossmann, Michele Angelo Besso, and Mileva Maric, the only woman in his class. There were the diversions of concerts and theater, as well as lively discussions at coffeehouses. Generally, Einstein found the scholarly life attractive, and as he neared the conclusion of his undergraduate studies he applied for an assistantship that would provide him with a stipend as he began work on a PhD. Although his grades were reasonably good, the assistantship was denied. He spent nine months seeking employment without success; then, in May 1901, he secured the first of two temporary positions, both teaching mathematics at technical high schools. Neither suited him.

In December he applied to the Swiss Patent Office in Bern, and in June 1902 he was hired as a "technical expert third class." Einstein later wrote that the greatest thing his friend Marcel Grossmann ever did for him was to recommend him for the position.[7] Here at last he found some measure of financial security and interesting work. In fact, he seems to have excelled at it: his supervisor called him "among the most esteemed experts at the office."[8] For Einstein, there was an important additional benefit. His friend Besso from the Polytechnikum was also employed there, and he was someone with whom Einstein could discuss physics.

So began an eventful period in Einstein's personal life. He married Mileva Maric in January 1903, and their first son, Hans Albert, was born in May of the following year. Einstein biographer Abraham Pais speaks of a remarkable and essential "apartness" in Einstein that enabled him to carry on his life immersed in thought, at the same time never being aloof from those around him or out of touch with world events. Despite the demands of a young family, in evenings and on weekends Einstein had the freedom to consider problems and compose articles. Over the next several years this is exactly what he did.

In 1905, Einstein produced six papers, several of which he placed in the prestigious *Annalen der Physik*. One concerned the light-quantum and photoelectric effect, which led to his Nobel Prize; one was a new determination of molecular dimensions, which was to become his doctoral thesis; and two were on Brownian motion, the random motion of microscopic particles suspended in a liquid or a gas. Any of these would have established his reputation as one of the most formidable intellects of his time, yet even they were overshadowed by the remaining two. These would change our understanding of the cosmos at a most fundamental level; both concerned what would come to be called *special relativity*.

The Aether

To appreciate special relativity, we should review some aspects of physics as they were understood in the waning years of the nineteenth century. The natural philosophers of that time were reasonably certain that a perfectly transparent, frictionless substance filled all of space. This was called *aether*. In the 1878 edition of the *Encyclopedia Britannica*, it was described by no less an authority than James Clerk Maxwell, the Scottish physicist whose 1873 *Treatise on Electricity and Magnetism* had introduced the equations that describe how electromagnetic energy propagates. The tone of his entry is remarkably assured: "There can be no doubt that the interplanetary and interstellar spaces are not empty but are occupied by a material substance or body, which is certainly the largest, and probably the most uniform body of which we have any knowledge."[9] It may strike us as a quaint idea, but we would do well to appreciate its rationale.

By the late nineteenth century, the debate over the nature of light—whether it was composed of particles or waves—was already centuries old. Newton had suspected that light was a particle (a "corpuscle" was his preferred term), but by the second half of the nineteenth century and the first years of the twentieth, most physicists believed it to be a wave. Physicists knew that, to be heard at all, a sound wave must travel through a medium—air, water, or a steel plate, for example. If light were also made of waves, they reasoned, then it, too, would require a sustaining medium.

It was about this time that the American physicist Albert A. Michelson suggested that Earth in its orbit would be moving through the aether, producing an "aether wind." He proposed that a measure of that wind would provide a means to determine the velocity of Earth moving through space. Much as one might gauge the speed of a river current by comparing the time it takes a swimmer to swim against the current and then at right angles to it, one might gauge the speed of Earth through the aether by measuring differences in the speed of light along the line of Earth's motion and at right angles to that motion.[10] In 1881, with a series of tests using an ingenious technique that has come to be called Michelson interferometry, he did just that. He found no evidence for an aether wind.

Six years later, Michelson conducted a more exacting version of the same experiment, this time in collaboration with Edward Williams Morley, a chemist from Western Reserve University. After several months they confirmed Michelson's result. Light had the same speed measured in all directions. It was a disturbing conclusion, among other reasons because it flew in the face of common sense. That is, if a train engine shines a headlight in a given direction when at rest, and shines the light again when it is moving in that direction, one might expect that an observer would measure two different speeds for the light. In the first instance it would be light's absolute speed, and in the second instance it would be its absolute speed added to the train engine's velocity. But Michelson and Morley had shown that light had a constant velocity no matter how fast the train ran—or in which direction.

Near the turn of the century, researchers were making some other rather startling findings. In 1898, Poincaré observed, "We have no direct intuition about the equality of two time intervals."[11] He questioned the concept of simultaneity and suggested that light might be a universal speed limit. The Irish physicist George Francis FitzGerald in 1889 suggested that the length of material bodies actually changes according to their velocity, and he explained Michelson and Morley's result by suggesting that the experimental apparatus, along with Earth itself, shrank in the direction of motion. Dutch physicist Hendrik Lorentz suggested that the electrical structure of matter causes it to contract when it is in motion, and that the contraction occurs along the direction of its motion. Finally, Lorentz and Poincaré had discovered a way to make Maxwell's electromagnetic laws simpler and more elegant, but they would have to assume that time itself flows more slowly when it is measured by someone moving than by someone at rest, and they were prepared to make no such claim. Without knowing it, they had come to the brink of special relativity, and stopped.

Special Relativity

That cognitive leap was made, of course, by Einstein, who described what would come to be called special relativity in his 1905 paper "On the Electrodynamics of Moving Bodies." To gain a sense of Einstein's insight, we must understand the features of *reference frames*—that is, imaginary frameworks in which bodies may be said to exist and with which they move.

Probably the first recorded experiment with reference frames was performed early in the seventeenth century by the French philosopher Pierre Gassendi. He dropped cannonballs from the mast of a ship, first when the ship was stationary and again when it was moving along a straight course at a constant speed. Each time, the cannonballs struck exactly the same place at the foot of the mast. From the ship, Gassendi measured the cannonball's total speed as the velocity of a falling object. From the dock he measured a greater speed: in addition to the velocity imparted by the fall was a sideways velocity imparted by the ship.

The discovery that light travels at the same speed regardless of the observer has intriguing consequences for this experiment. Let us repeat it, and for Gassendi's cannonball let us substitute a searchlight and a photocell. Suppose that Gassendi fixes the light to the top of the mast and points it directly at a photocell at the foot of the mast. Suppose he also has a device that can measure and record the interval between the moment we turn on the light and the moment its flash is detected by the photocell (Figure 2.1). Gassendi turns on the searchlight. The detector records that the light traveled from the top of the mast to the photocell at 299,792 kilometers a second. Since this is the value long established as the speed of light, all seems well.

Now suppose, however, that Gassendi's assistant is observing this experiment from the dock and Gassendi's ship is moving past him at ten knots. Following Einstein, we will simplify matters by assuming that the ship exists in a special sort of reference frame called an *inertial reference frame*—that is, a reference frame unaffected by external forces and driven solely by its own inertia. The ship will continue in the same uniform motion as it began, and we will discount all outside forces acting on it, including wind, friction from air and water, and gravity.

To Gassendi, the light beam travels along a vertical line from the top of the mast to the deck. But his assistant on the dock, as the ship moves steadily past him, sees the light travel a slightly longer, diagonal path—from the top of the mast when the ship was somewhere

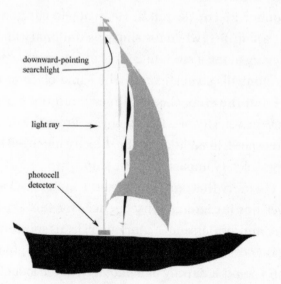

downward-pointing searchlight

light ray

photocell detector

Figure 2.1 Measuring the Speed of a Light Signal

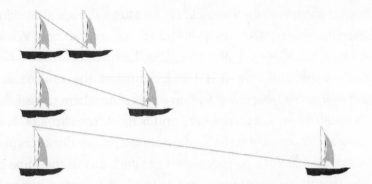

Figure 2.2 Three Light Signals Sent as Gassendi's Ship, Seen Here from His Assistant's Reference Frame, Moves at Three Different Speeds

Each left-hand silhouette represents the ship at the moment the signal is sent, and each right-hand silhouette represents the ship as the signal is received.

to his left, to the deck when the ship is somewhere to his right. Gassendi's assistant might expect to measure the total velocity of the light as 299,792 kilometers a second plus ten knots. Instead, he measures the speed as exactly 299,792 kilometers a second. Guessing that either his added speed was too small to be detected or that he made an error in his calculations, Gassendi makes the ship move faster and instructs his assistant to try again. His assistant measures the same results. Gassendi makes the ship move still faster, and there is still no change. In fact, even if he could somehow propel the ship at velocities within a whisker of the speed of light, his assistant would obtain the same result (Figure 2.2).

In each case, the assistant sees the light covering a greater distance than it did from Gassendi's perspective on the ship and in the same elapsed time that Gassendi measured on the ship. Speed equals distance over time. Clearly, something has to give. It cannot be speed, as light has one speed; it cannot travel faster than itself. It must be time and distance. And so it is. The assistant on the dock has no choice but to conclude that Gassendi's time on the ship is moving more slowly than is his time on the dock.

Now suppose that the assistant sets up a mast and a similar experimental apparatus on the dock. He measures the speed of light from searchlight to photocell at 299,792 kilometers a second. While passing the dock, Gassendi also measures the speed of light on his assistant's apparatus. He sees the light travel the longer, diagonal path—from the searchlight when it was somewhere to his left, to the photocell when it is somewhere on his right. The situation is exactly reversed. Gassendi sees the light covering a greater distance than it did from the assistant's perspective on the dock and in the same elapsed time that the assistant measured on the dock. Gassendi can only conclude that his assistant's time is moving more slowly than is his.

Who is right? The answer is that both are right. Special relativity claims that, in all inertial reference frames, the speed of light is the same. Moreover, it is the same in all directions, and its speed is unrelated to and independent of the observer. The theory further asserts that, in all inertial reference frames, the laws of physics are the same. This means that Newton's laws of motion and Maxwell's electromagnetic laws are valid in all inertial reference frames, no matter where those frames are or how fast they are moving. Finally, and perhaps most audaciously, special relativity says that the proper test of the validity of *any* law of physics is that it will be the same in every inertial reference frame.

Einstein was fully aware that his idea was radical, and he expected opposition from all quarters. What actually transpired was utter silence. The next issues of *Annalen der Physik* made no mention of his paper. Somewhat later, though, a letter arrived from Max Planck at the University of Berlin. Planck was one of the best-known physicists of the day, and he was asking for clarification on some points. Einstein was elated. In subsequent months, largely because of Planck's interest and support, special relativity became a legitimate topic of discussion and research.[12] By 1906, visitors had begun to arrive in Bern to discuss the theory with the man they were calling "Herr Professor Einstein."

In the next several years a number of scientists began thinking about the theory. Among these was the Russian-born, German-educated Hermann Minkowski, who had been Einstein's teacher in Zurich and who, as most biographies of Einstein note, had once referred to Einstein as "a lazy dog."[13] By 1908, Minkowski had revised his opinion rather dramatically upward. In September of that year, before the Eightieth Assembly of German Natural Scientists and Physicians, he delivered a presentation in which he introduced formulations and terms that would become indispensable tools for understanding and explaining Einstein's ideas.[14] In time it would become known as Minkowski's "Cologne Lecture." It begins with a ringing declaration: "The views of space and time which I wish to lay before you have sprung from the soil of experimental physics, and therein lies their strength. They are radical. Henceforth space by itself, and time by itself, are doomed to fade into mere shadows, and only a kind of union of the two will preserve an independent reality."[15]

In fact, we all have some sense of this reality—that is, a union of space and time—in our day-to-day lives. From the stands at a baseball game, we can see the ball struck before we hear the crack of the bat. This lag occurs, of course, because light travels much faster than sound—in fact, more than 880,000 times faster.[16] It is slowed slightly in passing through a transparent substance; air, for instance, slows it by about seventy kilometers per second. Although light travels as quickly as anything *can* travel, it is not instantaneous. Consequently, all signals, with all the information they contain, come to us at a delay.[17] It is for this reason that we can never know where anything *is*. All we can really know is where it was.[18]

Depicting Spacetime

Early in the print transcription of the Cologne Lecture, Minkowski remarks, "Nobody has ever noticed a place except at a time, or a time except at a place."[19] We might expand upon this to say that there can be no travel through time alone or space alone. All travels—including the hypothetical journeys described later in this book—are necessarily through spacetime.

Minkowski illustrated spacetime with what has come to be called a *Minkowski diagram* (Figure 2.3). As simple as it appears, the illustration has enormous explanatory power. Time is represented by the vertical axis, and because the three spatial dimensions are of a kind, they are represented (as "space or distance") by the single horizontal axis. In the center of the diagram, the "here and now" of Figure 2.3, is a point in space at a single moment. We term that point an *event*. Anything—a neutrino, a dandelion, a star, or a human being—might exist at that event. As it moves through time, it becomes what Hinton had called a filamentary atom and what Minkowski called a *worldline*.

A person who stays still—you sitting in a chair reading, for example—will move only along the time dimension and so will have a worldline that is vertical. If you get up and walk in a straight path at a steady speed of three miles an hour, your worldline will describe a greater angle from the vertical. If you walk somewhat

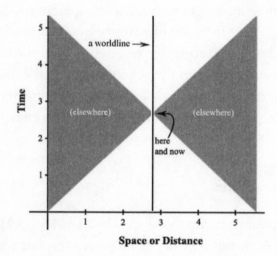

Figure 2.3 The Minkowski Diagram: A Worldline in Spacetime

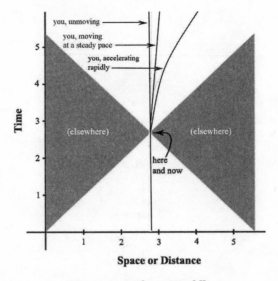

you, unmoving

you, moving
at a steady pace

you, accelerating
rapidly

(elsewhere)   (elsewhere)

here
and now

Time

Space or Distance

Figure 2.4 Three Worldlines

faster, your worldline will describe a still greater angle from the vertical. If you begin walking slowly, then walk faster, and then break into a run, your worldline will be of a body that accelerates. It will curve outward.

An ordinary spacetime diagram can accommodate minutes and miles, or hours and kilometers. Minkowski realized that, if he used units based upon the speed of light—say, hours on the time axis and light-hours on the space/distance axis—then the worldline of a photon moving at the speed of light in a vacuum would appear as a diagonal line at an angle of exactly forty-five degrees. (Realistically, the worldlines in Figure 2.4 would not diverge from the vertical as greatly as they do in the depiction.) Light is of course one of many kinds of electromagnetic radiation; all would travel along the diagonal.

We may give the diagram a second spatial dimension by rotating it along the vertical axis. When we do, the diagonals sweep out surfaces called *light cones* (Figure 2.5).[20] A signal or material body traveling at less than the speed of light may be sent from an event to

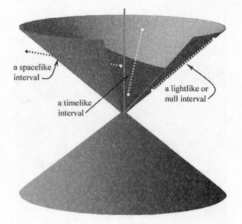

Figure 2.5 Future and Past Light Cones

any point inside its future light cone. The event and that point are separated by a *timelike interval*. A signal traveling at exactly the speed of light and sent from the same event travels along the surface of the light cone. It may be sent to any point on that surface. The event and that point are separated by a *lightlike*, or *null*, *interval*. Because the surface of the light cone represents the speed of light, a signal or a material body cannot be sent from an event to any point outside its light cone. Still, we may draw a line between the event and any such point. The event and that point are said to be separated be a *spacelike interval*.

The Myth of the Universal "Now"

Suppose that Gassendi wishes to send a signal at the speed of light to the vicinity of the star Arcturus, thirty-six light-years distant (Figure 2.6). A signal sent from the event that determines the light cone on Earth will take thirty-six years to reach the worldline of the event that determines the light cone of a point near Arcturus. Naturally, someone at that point (perhaps Gassendi's assistant) would have no way to know whether Gassendi sent a signal until it arrived.

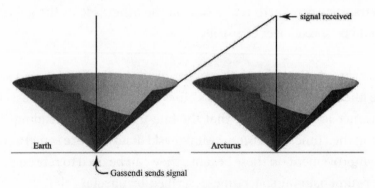

Figure 2.6 Sending a Signal to Arcturus

Here is where special relativity conflicts with our intuition of a single universal present. It says that there is no way to establish what occurred in the vicinity of Arcturus at the moment Gassendi sent the signal on Earth. We might expect that Gassendi's assistant could simply subtract the thirty-six years from the moment of the signal's arrival. But the thirty-six-year delay in itself is not the problem. Indeed, if Arcturus and Earth did not move relative to each other—that is, if they shared the same reference frame—subtracting the thirty-six years would yield a correct result. Like most bodies in space, however, Earth and Arcturus do move relative to each other. This means that clocks in the vicinity of Arcturus will tick at a different rate than clocks on Earth do. Both sets of clocks are correct, but they will differ as to how much time has elapsed since the signal was sent.

In fact, there was no single moment in the vicinity of Arcturus simultaneous with the moment the signal was sent from Earth. Rather, because of the time dilation effect of special relativity, there was a *range* of such moments. Between two points on Earth, this effect produces temporal ambiguities that amount to only small fractions of a second, and for most purposes they may be discounted. But the range of contending "nows" grows greater with distance. In the case of Earth and Arcturus, it would be a matter of hours. Between Earth and a star on the other side of the galaxy, it would be spread over centuries.

Illustrations that plot time against distance have been used since the Middle Ages. In fact Wells's Time Traveller refers his guests to a barometric chart, saying that the line traced by the readings was along the "time dimension." Minkowski diagrams are something of an improvement on these because they can be used to represent the four-dimensional spacetime described by special relativity. Compasses and protractors are perfectly suited to Euclidean space, but they are of little use for measuring space in higher dimensions. There is, though, a tool perfectly suited to such realms: it is coordinate geometry, which enables mathematicians to represent geometric forms with algebraic equations.[21]

To gain an intuitive sense of spacetime (and our deficiencies in perceiving it), we will forgo the mathematics in favor of an illustration. Imagine a four-walled room. On two adjacent walls are spotlights, each aimed at a screen on the wall opposite. Imagine that a right triangle is floating freely in the middle of the room, turning slowly on all three axes, and that the spotlights cast shadows of the triangle on the screens opposite. The dimensions of the "shadow triangles" shrink and grow continually, and at any given moment each shadow looks different from its counterpart on the adjacent wall. The shadows may be said to represent different two-dimensional slices of the single three-dimensional form that is the triangle.

So it is with our awareness of spacetime. Two observers in relative motion perceive different three-dimensional slices, or "shadows," of the deeper reality that is four-dimensional spacetime. Just as the lengths of the sides of the two shadow triangles are always different, so the spatial and temporal distances between any two events perceived by two observers in relative motion are different. Likewise, just as the lengths of the sides of the triangle in the center of the room are unchanging, so the distance between any two events in spacetime—the *spacetime interval*—is absolute. We cannot perceive that interval directly, but using a set of equations called the *Lorentz transformations*, we can measure it. Generally, we may say that the more a worldline leans or bends away from the vertical and toward the surface of the light cone, the shorter is the spacetime interval that it measures.[22] In the extreme case—that of an electromagnetic wave traveling along the surface of the light cone—the spacetime interval shrinks to zero. This means that, from the point of view of the electromagnetic wave, time is dilated infinitely (that is, time "stands still") and distance shrinks to nothing.

Futureward Time Travel

What, exactly, has all this to do with time travel? One implication of special relativity is that, as a body increases its velocity (whether circular motion or linear motion), it will experience *time dilation*; that is, its clock will tick more slowly than the clock of the universe outside.

Suppose that Gassendi has equipped his ship for travel through space and has installed in it an engine that can accelerate it to near-light velocities. Gassendi leaves his assistant at the dock on Earth and travels at near-light speed until he reaches the vicinity of Arcturus, thirty-six light-years from Earth. Immediately he slows, reverses direction, and travels again at near-light speed to return to Earth. When he arrives at the dock, he finds his assistant so much older that he barely recognizes him. In fact, the assistant has aged more than seventy-two years, while Gassendi has aged hardly at all. Gassendi may thus be said to have leaped ten years into the future.

Gassendi's assistant recognizes his master immediately and strikes the dock with his cane, saying that, although his faculties have dimmed, he clearly recalls learning that velocities are relative. From the point of view of Gassendi's ship, the assistant was the one doing the traveling. It seems to him that both he and Gassendi should have experienced an equal time dilation—that both should have aged eighteen years. At this, Gassendi, speaking compassionately although somewhat loudly, tells his assistant that the result he describes would indeed be the case if their relative motions had been symmetric. But they were not. His ship has not merely traveled *away* from Earth—it has also traveled *back*.

Were we to view the worldline of the assistant in his reference frame, we would see that it is vertical throughout its length, while Gassendi's worldline is bent away from the assistant's worldline during his outbound trip, and bent toward it during his inbound trip. But if we view the worldline of Gassendi in his own frame, we see that it is vertical only for the outbound leg; on the inbound leg his worldline must bend nearer the light cone to "catch up" with the worldline of his assistant. In both reference frames, Gassendi's worldline comes nearer the light cone than does the worldline of his assistant. Gassendi's time is dilated accordingly.

As strange as situations like this seem, there is nothing unusual about futureward time travel. Each of us is subject to a degree of time dilation every time we move; the faster we move, the greater the dilation. This is a very small degree, to be sure. Astronauts and cosmonauts on long-duration missions, orbiting Earth at seventeen thousand miles an hour for several months, are younger than they would be had they stayed on Earth by only a few fractions of a second. Although the effect is considerably less for those of us moving at more modest speeds, it is both real and inescapable. Futureward time travel, one might say, is an inevitable consequence of special relativity and of the universe in which we live.

Newton's Universe

Before the late seventeenth century, most Europeans held a conception of the cosmos that was essentially medieval. Earth was sur-

WHY SPECIAL RELATIVITY IS NOT INTUITIVE

Subatomic particles called mu mesons (or muons) may be created in particle accelerators, where they are known to have a decay rate of about two microseconds. They are also created naturally and continually as cosmic rays strike the upper atmosphere. At the moment of their creation, muons are traveling at very nearly the speed of light. Were it not for the time dilation effect of special relativity, then even at that great velocity most would travel less than a kilometer before they would decay, and all would have decayed at the upper fringes of the atmosphere. But as it happens, many muons are detected at Earth's surface. They survive the journey only because they experience time dilation.

Einstein's theory of special relativity is more than a century old, and it is confirmed on a daily basis. The word "relativity" and the phrase "everything is relative" have entered everyday speech and have drifted beyond physics and cosmology into other areas of human endeavor, like political theory and philosophy. For all this, most of us speak as though the universe ran according to an absolute time. We know better. And we also know better than to speak of the Sun rising, when we understand perfectly well that the Sun does not revolve around Earth. We recognize that what happens is that, as Earth rotates, our eastern horizon moves from the hemisphere of Earth in shadow into the hemisphere that is sunlit. It is true that we may place some blame on our language. English has no clear and economical expression to describe what really happens during a sunrise. But it is fair to ask why we have not—nearly five hundred years after Copernicus's *Revolutions of the Celestial Spheres*—developed such an expression. The reason may be simply that it would not describe our experience. From where we are standing, we see a Sun rising. If and until we regularly traverse interplanetary space and grow accustomed to thinking of planets as rotating spheres perpetually half-lit by their suns, we are likely to continue to speak of sunrises.

The same is true of time. To acknowledge relativistic effects in our communication would greatly complicate that communication. (Note that mentions of historical persons in this chapter placed them at points in space [like Zurich] and points in time [like 1905], not at events in spacetime.) But the other reason we neglect the effects of spacetime is simply that we move too slowly to be aware of them. In our day-to-day existence, differences between inertial reference frames are insignificant. In the equations of special relativity, the parameter that represents time is always multiplied by the speed of light, such that one second of time equals 299,792 kilometers of space. Even the greatest speeds that we can attain are so slow that the speed of light by comparison might as well be infinite. In short, for our everyday purposes, Newtonian ideas of time and space work quite well.

rounded by a celestial realm containing the Sun, the planets, and the stars, all held in their own spheres. The spheres themselves were enclosed within an outermost sphere called the empyrean, a place of purest fire imbued with the presence of God. If one could somehow move from the Earthly to the celestial, one would not find a transparent ceiling to break through. Rather, one would sense in one's surroundings a gradual diminution as the physical world faded gradually into the unphysical. This conception was a fiction (albeit an extraordinarily beautiful one) that most had accepted for centuries. Isaac Newton's 1687 *Principia* challenged it, arguing that the physical laws that governed events on Earth also governed the motions of the planets, and indeed, of everything in the celestial sphere. The falling apple, he maintained, was controlled by the same force that held the Moon in orbit.[23]

The *Principia* defines laws of motion and attraction. It explains mathematically the tides caused by the Sun and Moon, the flattening of Earth at the poles due to its rotation, the precession of its axis, and the orbits of the Moon, the planets, and comets. Perhaps most impressively, it makes predictions. As astronomers refined their measurements, those predictions were tested with ever-greater rigor. In 1845, French mathematician Urbain-Jean-Joseph

Leverrier suggested that an unknown body was causing a perturbation in the orbit of the planet Uranus, and he used Newton's laws to determine its probable location. He sent his calculations to Johann Gottfried Galle, an assistant at the Berlin Observatory. The evening of the very day he received the message, Galle focused his telescope on the suspect part of the sky and saw the planet that would be called Neptune.

The *Principia* is one of the most influential books ever written—not merely in physics or even science, but in any field. It has shaped our expectations and our beliefs. It is thanks to the *Principia*, for instance, that we regard the mysteries of the universe as solvable. Nonetheless, the work had limitations. Newton acknowledged that, although he had determined how celestial bodies and tides respond to gravity, he had not discovered the cause of gravity.[24] On this matter he would "frame no hypothesis" and insisted that it would be improper to do so. Einstein, however, was quite ready to frame hypotheses. In fact, his boldness on that score was nothing short of astonishing. With almost no empirical evidence he had dismissed Newton's conceptions of space and time; with even less evidence he would soon become so convinced of the correctness of general relativity that he believed that eventually the *Principia*, for all its magnificence, would be accommodated within it.[25]

Special relativity is "special" in the sense that it is a restricted case, operating in situations in which the effects of gravity and accelerations (that is, changes in speed and/or direction) may be discounted. Einstein's inertial reference frame was another such simplification. Although there are certainly regions of space where gravitational forces are balanced such that the total force acting on them may be said to be nil, these regions are so small that they are best measured on atomic scales. Sooner or later, even a dust particle in intergalactic space will be accelerated in one direction or another by the gravitational pull of galaxies. Einstein believed that special relativity would eventually be shown to be a special case of a more comprehensive theory, a natural law that would encompass *all* reference frames, including those influenced by accelerations and by gravity.[26]

The Principle of Equivalence

In 1906, Einstein gained a promotion at the patent office to technical expert second class, completed a paper on specific heats, and wrote a number of book reviews. Probably sometime in November of that year he experienced what he would call "the happiest thought of my life."[27] He recalled, "I was sitting in a chair in the patent office at Bern when all of a sudden a thought occurred to me: 'If a person falls freely he will not feel his own weight.' I was startled. This simple thought made a deep impression on me."[28] Einstein's insight was that gravity and acceleration are not separate phenomena that just happen to produce similar effects. They are "locally" equivalent; meaning that the laws of physics experienced in a small free-falling reference frame are the same as those experienced in a gravity-free inertial frame. This insight produced the postulate called the *principle of equivalence* or *equivalence principle*. As special relativity said that the laws of physics are the same in all *inertial* reference frames, so the equivalence principle said that they were also the same in all *free-falling* reference frames.

Suppose that Gassendi's assistant has an intense curiosity about natural laws that outweighs any concerns for the health of his master. He pushes Gassendi from the top of the mast, and at precisely the same moment he drops a cannonball so that it falls alongside him. For purposes of this thought experiment, let us discount the effects of atmospheric drag. If Gassendi looks at the cannonball and, in his mind's eye, dissolves all else around him, then he and the cannonball will seem (to him) to be suspended in space, far beyond the reach of any gravitational field.[29] Any experiments that he wishes to perform on the free-falling cannonball will produce precisely the same results as those of experiments conducted on the cannonball if it actually were beyond any gravitational field. Of course, Gassendi and the cannonball *are* falling. In fact, like all falling bodies near the surface of Earth, they are accelerating at a rate of 9.8 meters per second every second. Moreover, they *are* in a gravitational field—the best evidence of which is the ship's deck that, momentarily, both will meet.

Now suppose that a happily uninjured Gassendi and assistant (contrite but far more learned) are both standing on the deck of the ship, and it really *is* floating in space, far from any gravitational field. The ship is accelerated "upward"—that is, at right angles to its deck—at a rate of exactly 9.8 meters per second every second. Einstein's principle of equivalence asserts that, as long as there is no visual referent outside the ship (that is, nothing against which Gassendi and his assistant might measure their motion), there is no way they can determine that they are not in a uniform gravitational field. Not only are Newton's laws of motion the same; *all* physical laws are the same. Gassendi and his assistant on board the ship may perform all conceivable physical experiments in any and every branch of science; none of those experiments will enable them to determine whether they are accelerating or are in a uniform gravitational field.

Special relativity, which had operated in an idealized gravity-free realm, had been extended into the actual gravity-endowed universe. It was the first step toward what would be called the general theory of relativity.

Much like the inertial reference frame of special relativity, the equivalence principle involved an idealization. Thoroughly uniform gravitational fields do not exist. The strength of any real gravitational field diminishes over distance, and for this reason parts of a solid body moving through it will be subjected to a range of gravitational forces. These are called "tidal forces." As long as the forces that hold the body together are stronger than the tidal forces, the body will stay solid and its parts will move in accordance with their center of mass. But if the tidal forces are stronger, they will squeeze and stretch the body; and if they are strong enough, they will tear it apart.

Suppose that, having returned from space, Gassendi wishes to demonstrate the effects of tidal forces. Let us imagine that a tunnel extends from a place in the water just over the rail of Gassendi's ship all the way to the center of Earth. To avoid unnecessary complications, let us also imagine that Earth is not rotating. Gassendi, wishing to take experimental advantage of this remarkable topographical feature, holds his assistant over the rail, upside down by the ankles. The upside-down assistant, for his part, holds two cannonballs. The assistant releases the cannonballs at the same moment, and they fall into the tunnel side by side. Then Gassendi releases his assistant. As the assistant falls through the tunnel, he is pulled (very slightly) apart along his vertical axis—his head toward the center of Earth, his feet away from it. Meanwhile, he sees that the cannonballs, falling just ahead of him, are gradually drawn together until, at the center of Earth, they meet.

Special relativity ensures that any body moving in an inertial reference frame moves in accordance with Newton's first law of motion: in a straight line. The equivalence principle extends this idea to say that any body moving in a (small) free-falling reference frame also moves in a straight line. Here is where things get interesting. The cannonballs of Gassendi's assistant were moving in a free-falling reference frame, presumably along straight lines, and yet they *met*. Recall that, in Euclidean geometry, straight lines can never meet. When Einstein's reasoning reached a similar impasse, he understood that it could mean one of only two things. The first was that the equivalence principle is wrong, that bodies in free-falling reference frames do not travel in straight lines. This was a concession that Einstein was simply unwilling to make. The alternative—the *only* alternative—was that bodies in free-falling reference frames influenced by tidal gravity were traveling in straight lines, but a different *kind* of straight line, in a different kind of geometry. What made such lines meet was the shape of space itself.

General Relativity

If Einstein came upon special relativity with a sudden insight, the path he took to the theory that would come to be known as *general relativity* was torturous. He later wrote of "years of searching in the dark for a truth that one feels but cannot express."[30] He had understood that tidal gravity was caused by spacetime curvature, but he did not know exactly how spacetime curvature was produced. He asked Marcel Grossmann (who by this time was dean of the physics and mathematics department at the Zurich Polytechnikum) if there was a geometry that could be used to answer the question, and Grossmann responded that indeed there was: Riemannian geometry. In fact, by the first decade of the twentieth century there was a fairly substantial body of work in the field,[31] but it was in mathematics, not physics, and Einstein had not known of Riemann. He could not know that Riemann contributed to a mathematical legacy—and one solution to a mathematical mystery—that stretched back more than two thousand years.

In the third century BC, a mathematician and teacher known as Euclid of Alexandria codified the axioms of geometry in a textbook called *Elements*. The work identifies five axioms or postulates—that is, basic assertions that have been proven true—and from these deduces 465 theorems. It does this with such rigor that it was the standard to which Isaac Newton and Gottfried Wilhelm Leibniz aspired when, some nine centuries later, they invented calculus. Still, *Elements* contained a puzzle. Postulates one through four were easily demonstrable, and even obvious. There is one straight line connecting any two points, every straight line can be continued endlessly, and so on. But the fifth postulate, commonly called the "parallel postulate," was different, and not nearly as intuitive. It stated that, if we have a line and a point not on the line, we can draw only one line through the point that is also parallel to the line.[32] In the centuries following Euclid's death, mathematicians sought a proof for the postulate. The history of their search spans more than two thousand years and reaches across three continents.

The Mystery of the Parallels

Arab scholars were particularly intrigued by the parallel postulate. Over a course of four hundred years, at least five Arab mathematicians sought its proof—and failed.[33] Nonetheless, with commentary and a variety of new and ingenious approaches, they nourished and sustained its legacy. In the eighteenth century a Latin translation of an Arab work inspired the Jesuit mathematician Girolamo Cardano, whose own work on the problem encouraged the Swiss-German scientist and mathematician Johann Heinrich Lambert. Alas, neither found a proof.

In the early nineteenth century, Hungarian mathematician Farkas Bolyai attacked the problem with an intensity that approached obsession and left him despondent. He warned his son János, "I have traversed the bottomless night, which extinguished all light and joy in my life. I entreat you, leave the science of parallels alone."[34] But János was undeterred. His own inquiry into the deeper nature of Euclidean geometry became a source of excitement and profound satisfaction. He wrote his father, "I have made such wonderful discoveries that I have been almost overwhelmed by them, and it would be the cause of continual regret if they were lost. . . . In the meantime I can say only this: *I have created a new universe from nothing*."[35] In fact, János Bolyai was mapping an entirely new and self-consistent geometry. But he was not alone. The brilliant German mathematician Carl Friedrich Gauss said that not a year passed without someone writing a book on the problem of the parallels,[36] and his own unpublished notebooks reveal that in 1799 he had discovered what was essentially the same geometry that Bolyai found.[37] Meanwhile, several hundred miles to the east, a young mathematician named Nicolay Ivanovich Lobachevsky, also convinced that the parallel postulate could not be proved, was arriving at the same conclusions.

All three proved the validity of the parallel postulate and showed not only that it was necessary to Euclidean geometry, but that it was the single feature of Euclidean geometry that defined it *against* other geometries. In the process, each invented a self-consistent geometry in which all the axioms and postulates of Euclidean geometry are valid *except* the parallel postulate. This new geometry—now termed Lobachevsky geometry because it was Lobachevsky's work that had progressed the furthest—applies to a "hyperbolic" surface (that is, a surface shaped like a saddle with edges that extend to infinity). On such a surface, an infinite number of lines may be drawn through a point not on a line that are parallel to that line.

The search for a parallel postulate had shown that non-Euclidean geometries were possible and opened the door to another geometry, the one that Einstein would seek. This new geometry was developed by a twenty-eight-year-old German named Georg Friedrich Bernhard Riemann. In 1854, Riemann applied to Göttingen to teach as a privatdozent—that is, an instructor who could charge fees directly to his students. Candidates were required to deliver a provisionary lecture and to propose three topics in an order that reflected their own degree of preparedness. Riemann was familiar with his first and second subjects, but the third, despite its rather specific title—"On the Hypotheses Which Lie at the Foundation of Geometry"—was little more than a collection of notions. As it happened, the professor before whom Riemann was to deliver the talk was Gauss, by then in his seventies. Traditionally, the professor chose the first topic on the candidate's list. But Gauss found Riemann's last title especially intriguing. Thus Riemann, to his immense discomfort, was informed that this third topic would be the subject of his lecture.

Riemann's talk covered a great many topics, including the extension of geometry into higher dimensions that Minkowski would use to illustrate special relativity, as well as the suggestion that space itself might be spherical.[38] But most significantly, it described a way to measure the curvature of space. Some sixty years later, Einstein realized that Riemannian geometry was exactly the mathematical tool he needed.[39]

Special relativity posits that, although observers moving at different speeds will perceive space and time as stretching and shrinking, spacetime itself is rigid. General relativity, however, shows spacetime to be elastic—bent, warped, and stretched by gravity. Again we will forgo the mathematics for an illustration. The effect is

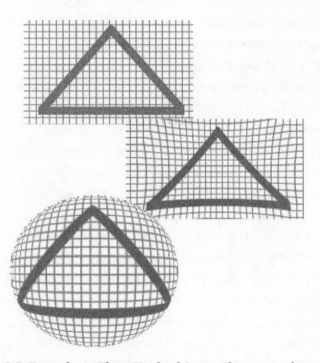

Figure 2.7 Triangles in Three Kinds of Space: (from top to bottom) Euclidean, Lobachevskian, and Riemannian

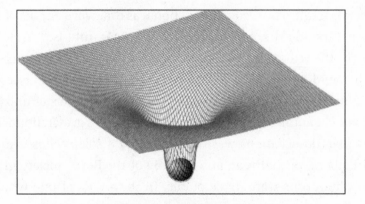

**Figure 2.8 An Embedding Diagram Depicting the
Warpage of Space by a Star**

best illustrated with an *embedding diagram*, in which space is represented by a two-dimensional elastic surface. By drawing grid lines
on this membrane we may represent *geodesic lines*, or *geodesics*—
that is, the shortest distances between points within this space, or if
we prefer, the paths of light rays (which, by definition, always trace
the shortest distance between two points). If we wish to represent
the effect upon spacetime of a massive body—say, a star—we drop
a sphere onto the membrane (Figure 2.8). The sphere sinks into it,
stretching and bending it into a shallow declivity. At some distance
from the sphere the geodesics still form a regular grid; the nearer
the body, the more the geodesics are stretched and the more the grid
is distended. The star's mass quite literally bends spacetime.

Einstein's Triumph

Einstein's theory of general relativity, which shows exactly how
mass bends spacetime, made three predictions. The first involved
a long-standing question regarding the planet Mercury. Mercury's closest approach to the Sun—its perihelion—shifted
slightly during each orbit, so that each perihelion occurred a bit
earlier in the orbit.[40] Einstein's result was in agreement with the

observed shift. The second prediction was that wavelengths of light from stars and the Sun would be *redshifted*—that is, "stretched" toward the red end of the spectrum—by gravity. But no such redshift could be verified. The third prediction, which carried the weight of a deciding vote, was that light emitted by a star and passing very near the Sun would be bent by the Sun's gravitational field. In 1919, observations were made during a solar eclipse, and on November 6 of that year, at a meeting of the Royal Society in London, the results were made public. In the words of one who was present, "The whole atmosphere of tense interest was exactly that of the Greek drama . . . There was dramatic quality in the very staging: the traditional ceremonial, and in the background the picture of Newton to remind us that the greatest of scientific generalizations was, now, after more than two centuries, to receive its first modification."[41] The evidence was decisively in favor of the value of displacement that had been predicted by Einstein.

Gravitational Time Dilation

Let us return to the second prediction, which, we should note, has since been verified repeatedly. Poke a stick in a pool of water. From the place of entry, a wave pattern generates outward in perfectly concentric circles. Drag that stick *through* the water, and this pattern changes. The waves generated ahead of the stick are pushed together, while those behind it are spread farther apart. This is a demonstration of the *Doppler effect* (or *Doppler shift*), and it operates for all manner of waves. Einstein's thoughts about the Doppler effect of light waves offered him an insight into a previously unexpected characteristic of gravity. As Einstein's reasoning was mostly mathematical, we will approximate it here with a thought experiment.[42]

Let us return again to Gassendi's ship. He fastens two searchlights to the mast—one at the top, another several meters below it. Both are pointed downward and are able to emit pulses of light at the same timed intervals. He releases the first searchlight and allows it to fall. As it begins its fall, it starts to emit light pulses. Immediately he releases the second searchlight, and it also begins to fall and to emit light pulses. Because the top searchlight was released first, it begins to accelerate first. Its speed therefore will be greater than that of the second searchlight for as long as both continue to fall.

Gassendi has equipped the second searchlight with a photocell and detector/recorder as before, so that it may receive light pulses from the first searchlight above it. The Doppler effect comes into play, and the light waves generated ahead of the first searchlight are pushed together. Consequently, the pulses that the second searchlight receives from the first searchlight are timed at a faster rate than its own. From this observation (in respectful imitation of Einstein) we may draw a rather startling conclusion: In the vicinity of the first searchlight at the top of the mast, time actually flows more quickly relative to time in the vicinity of the second search-light. In the vicinity of the second searchlight several meters below, time flows more slowly relative to time in the vicinity of the first. Extending the observation from the particular to the general, we may conclude that if a body (for instance, a searchlight) is at rest relative to a gravitational mass (for instance, Earth), the body's time will flow more slowly as it nears that mass. The phenomenon has come to be called "gravitational time dilation."

Like special relativity, general relativity provides a means to travel into the future—a consequence of the fact that time slows in the vicinity of gravitational fields and slows greatly in the vicinity of *intense* gravitational fields. The cases for which Einstein predicted gravitational time dilation would slow time by small amounts, detectable only by very precise measurements. But nature provides rather more extreme cases of the phenomenon—caused by masses so great that light itself cannot escape them. These are, of course, the collapsed stars called *black holes*. Black holes will play a greater role in later chapters, where they will be introduced properly. For now, though, let us define a black hole as a collapsed star having immense gravity, and let us define a black hole's *event horizon* as its "surface," the outermost edge of the roughly spherical region from which nothing—not even light—can escape.[43]

Black Holes and Time Dilation

A light cone in the vicinity of a sufficiently massive body is tilted toward that body, and the nearer the light cone is to the body, the greater the tilt becomes. Suppose the body is a black hole. We might imagine strings of light cones stretching from points in space to points very near a black hole's event horizon. Those farthest from the horizon would show only a slight tilt, those nearer the horizon somewhat more, until—very near the horizon—the leading edge of a light cone would subtend an angle a full ninety degrees from the vertical (Figure 2.9). Recall that the "vertical" dimension of a light cone represents time. Worldlines cannot be bent past forty-five degrees from the vertical with respect to their light cones. But suppose that the light cone itself is tilted. The worldline of a body within it would also be tilted, but the body would not experience time differently. If the light cone were tilted so that the worldline subtended an angle a full ninety degrees from the vertical, then relative to the universe outside, the body would be frozen in time.

A body falling toward a black hole would have the subjective sense that time was flowing at a normal rate, but someone watching from the outside universe would see the body moving ever more slowly, until—at the event horizon—it would stop moving altogether. (The Russians called black holes "frozen stars" for this reason.) For anything living, the fall would soon prove fatal, and inanimate matter falling beneath the event horizon could never be retrieved.

How might one use the effect to travel into the future? Hovering near the surface would be prohibitively expensive in fuel. But some-

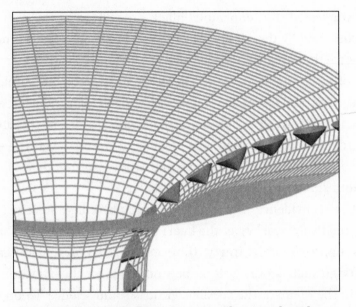

**Figure 2.9 An Embedding Diagram Showing the Tilting
of Light Cones in the Vicinity of a Black Hole**

thing in orbit around a black hole is *also* falling, not toward the
black hole but along a path around it. It would experience time dilation. Imagine, then, the occupant of a spacecraft in orbit around a
black hole. By choosing an appropriate altitude above the horizon,
he might determine the rate at which his own time passes. At a certain altitude a mere hour might pass while a year or a century
passed as measured in the outside universe. When he departed the
orbit and the vicinity of the black hole and returned to that universe, he would have traveled into the future.

Such an endeavor would face several complicating factors, one of
which would be tidal forces. For the types of black holes that physicists first imagined—that is, black holes between 2.5 and 100 solar
masses—gravity increases enormously over small distances above
the horizon, and tidal forces are extreme. The would-be time traveler would be torn apart before he experienced appreciable time
dilation. In 1969, however, British astrophysicist Donald Lynden-

Bell suggested that enormous black holes lived in the center of galaxies, and by the 1980s astronomers had begun to accumulate evidence that implied the existence of black holes with masses *millions* of times that of the Sun. The more massive the black hole, the weaker the tidal forces above its event horizon. (The tidal force of a black hole is directly proportional to its mass divided by the cube of its circumference.) It should be possible to get very near the event horizon of a massive black hole and enjoy extreme time dilation without suffering extreme tidal forces.

There is evidence that a massive black hole lives nearby (or at least relatively nearby), at the heart of our own Milky Way Galaxy. It is approximately thirty thousand light-years from Earth. Although such a journey is far beyond our present capabilities, it is within the realm of the possible. Futurists and science fiction writers have long imagined a variety of means by which interstellar reaches might be crossed—among them, various types of metabolic suspension; slow moving "space arks," in which generations beget generations; automated spacecraft carrying human ova that are inseminated at the appropriate moment in the journey, so in due course produce humans who are raised and tutored by onboard computers; and travel at near-light velocities that allow passengers to experience time dilation en route. In practical terms, none are ideal. Even the last—the fastest means described—would require a voyage lasting nearly thirty thousand years as measured on Earth.

For more convenient futureward travel of this type, we would need a supermassive black hole in our neighborhood. We might imagine a sufficiently advanced civilization undertaking a vast construction project in which stars are judiciously nudged from their courses toward a common point in space, where they collide and merge, one after another, to create a star with greater and greater mass. When it achieves a critical mass, the star collapses into a small black hole into which more stars are "fed," until it grows to such size that tidal forces above its event horizon are survivable. This black hole would be very massive indeed. Calculations by Misner, Thorne, and Wheeler suggest that it would weigh more than ten thousand Suns and so have a circumference of a hundred thousand kilometers.[44] Of course, such a construction project would come at considerable cost in materials—at least ten thousand average-sized stars or their equivalent. This number is put in perspective when we appreciate that roughly six thousand individual stars are visible to the naked eye in a clear, dark sky. Our hypothetical engineers would have to empty vast regions of Earth's heavens.

Mass Shells

The skill to manipulate and shape somewhat smaller masses (say, those of a large planet or small star) might be all we need to fashion another means to travel futureward. Newton had considered the gravitational properties of a spherical envelope of dense matter that physicists call a *mass shell*, and he arrived at some interesting conclusions. A person *outside* the shell, whether falling toward it or standing on its surface, would feel its gravity as though it were a solid sphere—that is, as though its gravity were emanating from a point at its center. But a person in the space inside the shell would have quite a different experience. He would be pulled outward in all directions. Oddly, the sensation would not be uncomfortable. Because the gravitational attraction between masses is inversely proportional to the square of the distance between them, all the gravitational forces acting on him would cancel each other, and any tidal effects would likewise be nullified. A person anywhere in the space inside the shell would be pleasantly weightless.

Cornell astronomer Thomas Gold was widely known as the man who, in 1968, had first suggested that pulsars might be neutron stars.[45] Few knew that, some years later, he gauged a mass shell's gravitational properties using Einstein's understanding of gravity (that is, as a warpage of spacetime) and arrived at a result identical to Newton's, except for an additional feature—one that, for our purposes, is rather significant: in the space enclosed by a mass shell, a clock would flow at a rate slower than the space outside the shell. To dramatize his conclusion, Gold introduced a parable:

> A mother, desiring her baby to remain young for as long as possible, decided on the following plan of action. Each night, when the baby was asleep, she went out and collected masses from afar, symmetrically, and arranged them in a spherical shell around the baby's crib (herself staying outside and somewhat away from the completed shell). The baby's sleep was not disturbed, no acceleration being measurable inside the shell. In the morning before the baby woke up, she dismantled the shell and stored the masses far away.[46]

The mother could slow the flow of time inside the shell either by adding mass to the shell or by shrinking its diameter. She would require a delicate touch: a shell that would yield an appreciable time dilation would have a diameter very near the threshold for gravitational collapse.[47]

In his 2001 book *Time Travel in Einstein's Universe*, Princeton astrophysicist J. Richard Gott imagines such a shell. It might be made, he suggests, by dismantling the planet Jupiter and rebuilding it. Although the mass of the planet Jupiter is considerable, it is a small fraction of that required to create the massive black hole described earlier. Perhaps accordingly, the degree of time dilation is likewise modest. The hollow space inside would be six meters across, and the flow of time there would be slowed by a factor of four. This would mean that the baby of Gold's parable, if kept inside continuously, would age three months for every year his mother aged. If the shell were dismantled for twelve hours of every twenty-four, the baby would age six months for every year she aged.

In sum, then, we have three ways to effect appreciable time dilation, or what we may call time travel to the future: linear motion outbound and inbound at relativistic velocities, circular motion at relativistic velocities, and subjection to a strong gravitational field like that in the vicinity of a black hole or inside a mass shell. Travel into the past, however, is a far more challenging proposition. Pastward time travel, if possible at all, would involve special relativity and general relativity in combination. It would also require vast energies and unimagined technologies, and might be bedeviled with logical paradoxes. Nonetheless, building upon the implications of Einstein's theories, in the 1920s and then in the late 1970s, two physicists and one mathematician made tentative inquiries in that direction.

CHAPTER THREE

"UNPHYSICAL" TIME MACHINES

*Is it not possible—I often wonder—that things
we have felt with great intensity have an existence
independent of our minds; are in fact still in
existence? And if so, will it not be possible, in
time, that some device will be invented by which
we can tap them? . . . Instead of remembering
here a scene and there a sound, I shall fit a plug
into the wall; and listen in to the past.*

— Virginia Woolf, "A Sketch of the Past"

As I write this chapter, an experiment in high Earth orbit is testing an effect of general relativity with unprecedented accuracy. The experiment is aboard a NASA spacecraft with the rather prosaic name *Gravity Probe B*. Its mission is to measure a phenomenon predicted nearly a century ago. In 1918, three years after Einstein showed that mass bends or warps spacetime, Austrian physicists Joseph Lense and Hans Thirring found that if the mass is spinning, it makes the spacetime around it spin too. The phenomenon has rough analogues that are familiar to us. A spindle rotating in a thick liquid, for instance, will pull the liquid around it into a swirl. Replacing the spindle with a sufficiently massive body and the liquid with spacetime gives the general sense of the phenomenon that Lense and Thirring were describing. It is sometimes called, appropriately enough, the Lense-Thirring effect, but it is more commonly referred to as frame dragging, because inertial reference frames are being pulled or dragged around the spinning mass.

The experiment on *Gravity Probe B* is designed to measure the frame dragging of Earth's rotation, and it centers upon four rapidly spinning quartz spheres, roughly the size of Ping-Pong balls. They may well be the most perfect orbs ever made: if Earth were as smooth, Mount Everest would have an altitude of six and a half feet. When the experiment began, they were precisely aligned with the star IM Pegasi, and as it proceeds, the spheres—spinning at ten thousand rotations per minute and acting as gyroscopes—should begin to tilt slightly in the direction of Earth's rotation. Because the tilt is expected to be quite small (it will be difficult to detect, let alone measure), the instrument is calibrated with a precision thirty million times more exacting than any gyroscope in existence. The smallest change in the spheres' molecular structure would upset their spin, so they are held in a thermos chilled to two degrees above absolute zero. The thermos is contained within a chamber that shields it from sound waves and magnetic fields, and the chamber floats freely in a compartment within the spacecraft. Finally, the spacecraft itself senses stray wisps of atmosphere and continually adjusts its orientation to compensate. By these means the four spheres are protected from every imaginable disturbing force— except, of course, the one they are designed to detect.

We may visualize frame dragging by imagining its effect on light cones. As a massive body will tilt a light cone toward it, so a massive *rotating* body, causing spacetime to swirl around it, will tilt a light cone "sideways" in such a manner that its long axis is inclined toward the direction of the body's rotation. The faster the rotation or the nearer the light cone to the body, the greater the light cone's degree of tilt.

Van Stockum

An effect of tilted light cones was first described in 1937 by a twenty-seven-year-old Dutch-born physicist named Willem Jacob van Stockum. He had been born in Hatten, the Netherlands, to parents of some distinction. His father, a first cousin of Vincent van Gogh, had served as a captain in the Dutch Royal Navy.[1] The family moved to Ireland when Willem was a child, and at age nineteen he began studies in mathematics at Trinity College in Dublin. By the mid 1930s he was preparing for his PhD at the Mathematical Institute at the University of Edinburgh. He hoped to work with Einstein at the Institute for Advanced Study and eventually gain a teaching position in the United States.

When war was declared upon Germany in June 1941, however, Van Stockum put all such plans aside and began training as a pilot with the Royal Canadian Air Force. There, in a testing and development division, his scientific background was put to good use; but he had a hunger for experience and a desire to make a more direct contribution, so early in 1943 he transferred to the Dutch contingent of the Royal Air Force as a bomber pilot. Soon Van Stockum and his crew were based in Yorkshire and flying missions over occupied Europe, including one in advance of the D-day invasion. On the night of June 10, 1944, during Van Stockum's sixth mission, his plane was hit with antiaircraft fire and fatally disabled. It crashed in the French countryside near the village of Entrammes. There were no survivors.

Van Stockum's written legacy rests with two pieces: a clear-eyed manifesto discussing the reason a soldier fights in a war,[2] and a technical paper published in *Proceedings of the Royal Society of Edinburgh* in 1937.[3]

The Van Stockum Cylinder

The paper in which Einstein formulated general relativity contains sixteen equations that comprise the fundamental formulas for describing gravitational effects. These are called *Einstein's field equations*, and they may be solved for any imaginable mass. In 1915, Karl Schwarzschild had solved the equations for a mathematical point and a sphere. Other physicists followed suit, and by the mid 1930s the equations had been solved for a variety of bodies—some realistic, others fully imaginary.

Van Stockum's paper solved the equations for a body in the second category: an infinitely long cylinder of rapidly rotating dust. The dust particles would be held out against their own gravity by centrifugal forces, and the cylinder's rotation would drag inertial frames so strongly that light cones all along its outer "surface" would tilt forward in the direction of that rotation.

Suppose that we have two "untilted" light cones associated with two events in spacetime: Gassendi sending a signal and his assistant sending a signal. Suppose also that these events are separated by several light-years. Over time, of course, Gassendi will receive the signal sent by his assistant, and the assistant will receive the signal sent by Gassendi. But at the moment each sends the signal, he can neither influence nor be influenced by the other, because any such influence (whether electromagnetic wave or material particle) would have to travel outside its light cone—an impossibility. Now, however, suppose we imagine that these same two adjacent light-cone pairs are positioned along the surface of Van Stockum's cylinder, and tilted by frame dragging. If they are tilted enough, the future light cone of the "trailing" light-cone pair will intersect the past light cone of the "leading" light-cone pair. Thus a signal sent by Gassendi might never travel outside its light cone yet still reach the assistant before the assistant sends his signal (Figure 3.1).

Van Stockum noted that an observer at rest on that surface sent a

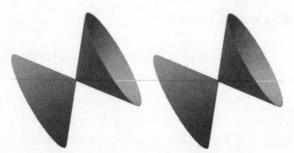

Figure 3.1 Two Pairs of Tilted Light Cones

light signal along it, frame dragging would warp it completely around the cylinder. The signal would be said to travel along a *closed null curve*—"closed" because it meets itself, and "null" because its length in spacetime is zero. The implications were remarkable. An observer on the "surface" of the cylinder would be able to see entirely around it.

Because the cylinder in question was infinitely long, and nothing in our universe can be infinitely long, the whole scenario fell into a category that a physicist politely terms unphysical. It was for this reason that, as provocative as the paper was, in subsequent years it was cited rarely, and then only among lists of solutions to the field equations. Van Stockum himself was mostly forgotten. In fact, until quite recently few knew that the author of the article in question was also the bomber pilot.

Although Van Stockum seemed not to have realized it, his paper described a means to send signals (and implied a means to send material bodies) pastward.[4] If, at a given altitude above the cylinder, light cones were tilted such that a signal might travel along their surfaces through a closed null curve, it followed that, at a lower altitude, light cones would be tilted still further, enough that one might travel through their interiors through a closed timelike curve— "closed" (again) because it meets itself and "timelike" because a material body could travel the length of the curve at less than the speed of light.

Gödel

By the time Einstein was in residence at the Institute for Advanced Study, his celebrity in some ways insulated him from other scientists, perhaps *especially* other scientists, as it was they who fully appreciated his achievements. He rarely attended a lecture at Palmer Physical Laboratory at Princeton, but when he did he caused an "awed hush."[5] He wrote to a friend, "I am generally regarded as a sort of petrified object, rendered blind and deaf by the years. I find this role not too distasteful, as it corresponds very well with my temperament."[6] He was exaggerating considerably; his grace and self-effacing humor were unfailing. Yet by the 1950s, Einstein's famous skepticism regarding quantum theory had cast him in the role of outsider, and it is not difficult to appreciate Einstein's longing for the company of an intellect that would not defer to him. In fact, there was such an intellect, in the person of mathematician and logician Kurt Gödel.[7] In his later career, Einstein once said that his own work meant little, and that he visited the institute building only to have the privilege of walking home with Gödel.[8]

Kurt Gödel was born in 1906 in Brno, Moravia. As a child, he asked so many questions about the world, and asked them so often, that his family called him *Herr Warum* ("Mr. Why"). By his teens he had developed interests in mathematics and philosophy, and in 1924 he began study at the University of Vienna. There he became involved with a like-minded group of philosophers who would be known as the Vienna Circle. Their company made for especially stimulating years. A friend recalled, "We met frequently for walks through the parks of Vienna, and of course in cafés had endless discussions . . . sometimes deep into the night."[9]

Gödel entered the university with plans to work in theoretical physics, but gradually he became more and more intrigued by mathematics, and when he began doctoral work (still at the University of Vienna) he made it his preferred field of study. His thesis, submitted in 1930, was a "completeness proof," a verification that logical values covered all they were meant to. It was a significant work, but it was no breakthrough; quite the contrary, it was an affirmation of the status quo. In the same year, though, Gödel produced another work, far more important and—to the mathematical establishment at least—far more disturbing.

The paper took as its subject elementary number theory—that is, the fundamentals concerned with the addition and multiplication of whole numbers, and demonstrated that even in this modest realm there are truths, some as simple as $2 + 2 = 4$, that are utterly unprovable. The work, which came to be known as Gödel's incompleteness theorem, upset the most fundamental of tenets and shook mathematics to its core.

It was a version of the Epimenides paradox, or "liar's paradox," whose clearest and simplest formulation may be in the assertion, "This statement is false." However we approach it—by assuming it is true or assuming it is false—we are immediately forced into the contrary reading. That reading, in turn, forces us back to the first reading, which forces us to return to the second reading, and so on—like an endless series of receding images in facing mirrors. The incompleteness theorem, despite its evidently negative character, produced constructive results. In time, the means by which Gödel achieved the proof generated recursion theory, a branch of mathematics that would become central to computer design.

When the Nazis invaded Austria in 1938, Gödel's personal situation became intolerable. In 1940 he and his wife Adele immigrated to the United States and made Princeton their home. Although Gödel had met Einstein years earlier, their new proximity engendered a lasting friendship. In some ways it was a curious relationship, in that, as Einstein's assistant Ernst Gabor Strauss observed, they were very different men: "Einstein [was] gregarious, happy, full of laughter and common sense, and Gödel extremely solemn, very serious, quite solitary, and distrustful of common sense as a means of arriving at the truth."[10] Yet they had a great deal in common. Both had German as a first language; both had benefited from the rich intellectual life of European universities; and both had accomplished their most significant work at early ages. Moreover, each knew enough of the other's field to appreciate his work within it, but not so much that he cared to compete. There was a still more fundamental similarity: both men, Strauss said, "went directly and wholeheartedly to the questions at the very center of things."[11]

Gödel and Einstein discussed their work almost daily, and there was much work to discuss. Gödel made major contributions to set theory, the mathematics of infinite classes, while Einstein was increasingly concerned with unifying gravitation and electromagnetism. They also talked of philosophy and politics. Strauss reports that one day in 1953 Einstein said to him, "You know, Gödel has gone completely crazy." When Strauss asked, "Well what worse could he have done?" Einstein said, "He voted for Eisenhower."[12] One cannot help but wish to have been a few steps behind them, but within earshot, during those walks home from the institute.

In early 1949, when Gödel was in his early forties, he wrote, "March 14 is Einstein's seventieth birthday and I don't know what I should give him. So far as I know him, he is not at all fond of such things. Adele knit for him a wool vest. . . . After long searches I finally sent him an etching."[13] But in fact, Gödel had already crafted another present: a five-page essay contributed to a collection being edited in honor of the occasion.[14] It claimed that ideas put forth by the philosopher Immanuel Kant and those presented by general relativity both suggested that time and change have no objective reality and may be nothing more than human inventions. Initially, Gödel had intended a more extended work, but as the writing progressed he decided that the short piece would suffice, and he resolved to treat its physical implications in detail in a separate paper. Indeed, those implications merited a distinct platform, as they were fairly startling. Gödel had discovered a new solution to the field equations—one that allowed for time travel to the past.

Gödel's Rotating Universe

In 1949, Gödel suggested that tilted light cones might occur naturally if the entire universe were both "static"—that is, neither expanding nor contracting—and in a state of uniform rotation. This universe would not rotate around any single point. But if we chose a point arbitrarily as the origin of a coordinate system and moved away from that origin, we would find that the tilt of light cones gradually increased until, at a certain distance, we would encounter a great ring of light cones stretching around the universe, each one tilted so far forward that the future light cone of a given event overlapped the past light cone of another. It would be possible to send a light signal through all these light cones.

Because to an outside observer the path of this signal would be nearly horizontal, it would display rather peculiar behavior. In Gödel's words, "the light signal will come back at exactly the same moment at which it is sent."[15] Moreover, he observed, "Since a light path can be approximated as closely as you wish by a path of a material particle, you can even travel into the past on a rocket ship of sufficiently high velocity."[16]

Gödel was aware that his hypothetical ring of light cones would be enormous. He estimated that its radius would approximate that of the "Einstein static universe"—that is, somewhere between ten and a hundred billion light-years. Accordingly, Gödel acknowledged that any test of his hypothesis would be impractical.

By 1949, astronomers had demonstrated beyond a reasonable doubt that the universe was not static; it was expanding.[17] They had also shown that it had no rotation. So by the second half of the twentieth century, Gödel's rotating universe was regarded as a rather quaint intellectual exercise, like a speculation on the feeding habits of unicorns. Physicists cited his "rotating universe" only occasionally, and Gödel himself published no further work on the subject.

Nonetheless, Gödel's work had significance. It showed that past-ward time travel was not prohibited by any known law of nature. If it was prohibited at all, it was by two rather more parochial circumstances. One was the limitations of human technology in 1949—and indeed the limitations at least several decades beyond, as no means of rapid transport across billions of light years is likely to be developed anytime soon. The other was a set of governing factors that, by sheer happenstance, denied the universe a rotation: the universe, it seems, might as easily have been rotating as not. In the larger view—that is, what nature itself allows and does not allow—both were trivial impediments. Einstein himself admitted that he had been troubled by the implications of closed timelike curves as he developed the general theory of relativity, but he had not clarified the issue as had Gödel.[18] It was clear that general relativity, the most comprehensive understanding of the large-scale structure of the universe, permitted time travel to the past.

Despite Gödel's modest estimation of the significance of his work in 1949, it seems that, in time, he came to consider it anything but trivial. The editors of the reference works generically titled "Who's Who" solicit their potted biographies from the subjects themselves. In *International Who's Who 1971–72*, Gödel chose to list only a handful of publications, and his work on a rotating universe was among them.[19]

Like Einstein, Gödel was somewhat isolated at Princeton but also had admirers. The physicist John Archibald Wheeler, who was much interested in the philosophical aspects of the study of the universe, had heard Gödel's 1949 talk at Princeton. Some twenty years later, when he was collaborating on a textbook on gravitation, he and his coauthors paid a visit to Gödel, then in his sixties, at his office at the institute. After introductions, Wheeler informed Gödel of their work. When Gödel asked the visitors what they intended to say about his rotating universe, they admitted that they had no plans to mention it. Wheeler said "the response distressed him."[20]

Sometime later, Wheeler learned that Gödel had not given up on the idea; quite the contrary, he had sought to prove it observationally. He could not measure the universe's rotation against anything outside the universe, as there is no such place. But he could measure it as *Gravity Probe B* gauges frame dragging around Earth—that is, with a freely spinning gyroscope. Gödel had no *Gravity Probe B*, but there was another way: the galaxies *themselves* might act as gyroscopes. If the universe were rotating, Gödel reasoned, the axes of rotation of individual galaxies would not be entirely random, but would tend toward an axis in the direction of that larger rotation. Wheeler learned that Gödel had worked through the *Hubble Atlas of Galaxies*, seeking a preferred axis of rotation, but had been unable to find one.

It is tempting to speculate on what sustained such an interest. Throughout his life, Gödel was beset with illness, and as he grew older he became increasingly—some said obsessively—concerned with his health. In fact, he was not made a professor at the institute until 1953, thirteen years after his residency began, because the institute's mathematics faculty worried that his paranoia and fear of making decisions were unsuited to the responsibilities associated with the position. In his own autobiography, Wheeler indulged in a bit of armchair psychoanalysis. He wrote that, in Gödel's rotating universe one could, in principle, live one's life over, and he concluded, "Gödel's passionate concern for his own health, so openly visible, was matched by an equally passionate, if less visible, wish to defy death and live again."[21]

Wheeler did not speculate on exactly how the Gödel universe would allow such a wish to be realized. Certainly there could be no rebirth in the literal senses of that word. Gödel's aging could not be reversed; he could not suddenly become a child again. What the ring of light cones *could* offer would be a means by which Gödel could visit his own past. If he could reach the tilted light cones and travel through them around the universe until he reached his own worldline in, say, 1912, then a sixty-year-old Gödel could meet, converse with, and answer the questions of a six-year-old Mr. Why.

Gödel's complexities do not give themselves up to easy analysis, and at any rate, our concerns are elsewhere. Nonetheless, it is difficult to completely disregard what is at least an oblique connection between the pastward time-traveling possibilities allowed by Gödel's rotating universe and what may be his conception of an afterlife. In 1961 he composed several letters to his mother in which he insisted that the world is arranged in a rational and meaningful way, and that lifetimes are too short for the proper exploration of the many possible courses that an individual might take.[22] It followed, he continued, that there must be a subsequent existence in which humans can better explore such paths and fully realize their potentials. Whether he was expressing his own beliefs or merely comforting a woman facing her mortality, we cannot know. But let us suppose it was the first. How might such ideas relate to the rotating universe? One might imagine a time traveler advising and redirecting his younger self to avoid errors that he had made on the first circuit. But this is speculation. Gödel himself seems never to have connected his ideas of afterlives to time travel.

Before we return to firmer ground, we might mention another admittedly slender thread by which Gödel might be linked with pastward time travel, or a most interesting aspect of it. Specifically, the endless self-negations of the incompleteness theorem rather nicely reflect time travel's central and most famous paradox.[23] Suppose that the sixty-year-old Gödel meets the six-year-old Mr. Why in 1912 and takes some action to ensure that he never becomes the sixty-year-old who makes the trip through the light cones. He might persuade his younger self to adopt a course of study in, say, art history, in which he would never learn to solve the field equations, and so never discover the means by which he might make the trip. Or the elder Gödel could employ a more radical course of action that would produce far more certain results. He could simply kill his younger self. The action gives rise to the paradox in a cleaner form. If Gödel dies at age six, then he cannot live to age sixty. If he does not live to age sixty, there is no one to go back in time and kill the younger self. If there is no one to go back in time and kill the younger self, then the younger self will live to age sixty. If the younger self lives to age sixty, then he will go back in time. And so on, in an endless causal loop.

Tipler

Theoretical physics is a vast and complex terrain. Some parts are long settled and familiar; others are newly mapped. But there remain regions that have only vaguely discernible outlines and may not be there at all. Among the physicists who think about those places is Frank Tipler. "Good scientists," he told an interviewer in 1994, "have chutzpah."[24] In fact, Tipler has chutzpah to burn, and it has been evident since childhood. In the early 1950s, Wernher von Braun, then the world's best-known rocket and aerospace scientist, announced plans for launching an artificial Earth satellite. A kindergarten-age Tipler wrote him a letter requesting rocket fuel. Tipler never received the fuel, but in subsequent years he would manage a good deal of space exploration without it. He earned an undergraduate degree from MIT, and by 1974 he was a twenty-seven-year-old PhD candidate in physics at the University of Maryland.

The Tipler Cylinder

As part of his graduate work, Tipler imagined an experiment that could be conducted only "in principle" because the engineering it required was quite beyond human abilities, and likely to remain so for some time. In the April 1974 issue of *Physical Review D*, Tipler suggested that if a cylinder had a sufficiently powerful gravitational field and a sufficiently fast rotation, it would drag spacetime around it in the direction of its rotation. He credited the solutions offered in Van Stockum's 1937 article. He also acknowledged Gödel; indeed, the cylinder might be thought of as a small part of the collective mass of Gödel's universe compacted to a manageable size.

To elucidate, let us call upon our intrepid seventeenth-century experimenters to demonstrate another lesson in reference frames, perhaps the strangest so far. Suppose that Gassendi and his assistant, each in his own ship, approach a massive rotating cylinder somewhere in deep space. At a place where the gravitational pull of the cylinder is still unfelt, Gassendi stops his ship and bids farewell to his assistant. The assistant keeps his ship on its course, allowing it gradually to come under the influence of the cylinder's gravity. Soon his ship is pulled around the cylinder in the direction of its rotation, and the nearer the ship gets to the surface of the cylinder, the faster it is carried around it.

The assistant notices nothing unusual about his perception of time. The dials on the onboard chronometer move as they always have; his pulse beats at the same rate it always has. Likewise, Gassendi, standing on the deck of his ship beyond the reach of the frame-dragging effect, perceives no change in his perception of time in his immediate vicinity. But if the assistant, as he neared the cylinder, could send Gassendi signals—by, say, radio waves— then Gassendi would detect them arriving at a slower and slower rate. If Gassendi could view the deck of his assistant's ship through a sufficiently powerful telescope, he would see his assistant moving ever more slowly.

Imagine a two-dimensional disk, like a vinyl record impaled by a spindle. Like the grooves in the record, there are circular orbits within the plane and around the cylinder. Future light cones projected from events in outer orbits would be nearly vertical, but future light cones projected from events in nearer orbits would be tilted in the direction of the cylinder's rotation, and the nearer they were to the cylinder, the greater would be their tilt—until future light cones at the surface of the cylinder would have a central axis that lay very near the plane or completely within it.

As Gassendi's assistant nears the cylinder, his future light cone becomes more and more tilted. Of course, one cannot sense the tilt of a light cone from inside it, no matter how much it tilts. Even in a greatly tilted light cone, one would sense that one's worldline, so long as one remained still, was vertical. Thus, Gassendi's assistant would sense no change in the flow of time aboard his ship. But if Gassendi could have seen a spacetime representation of his assistant's journey to the cylinder, it might look like Figure 3.2.

The greater the tilt of a light cone, the more slowly everything within it moves when viewed from an outside reference frame. Suppose that, from Gassendi's reference frame, the light cone of his assistant's ship is tilted so far forward that the worldline through its center becomes horizontal. When Gassendi observes the ship's deck through his telescope, he sees no motion whatsoever. The chronometer on the assistant's ship has stopped; the assistant himself appears as a statue.

Figure 3.2 Frame Dragging around a Massive Rotating Cylinder

For his part, the assistant would be at a very strange point in spacetime. The lower half of his future light cone would lie below the horizontal plane. Any signal he sent into that region would travel into the past, and indeed if Gassendi's assistant took his ship into that region, then *he* would be traveling into the past. (But even if the ship were not so near the cylinder and its light cone were not tilted so extremely, signals to the past and time travel to the past would still be possible. If, for instance, the ship's worldline were tilted only a shade more than 22.5 degrees from the vertical, enough so that the leading edge of the light cone dipped slightly below the horizontal plane, then the assistant would be able to send a signal at the speed of light along that surface.)

If the assistant chose to travel pastward in time, then he could choose the distance traveled by the angle that his worldline subtended below the horizontal plane (the greater the angle, the farther into the past he would travel) and/or the number of orbits around the cylinder (the more orbits, the farther into the past he would travel). Thus, Gassendi's assistant might, by careful navigation down a sort of spacetime winding staircase, reach a moment a few hours earlier, when Gassendi bid him farewell. He could take a few more turns around and arrive at whatever day, year, or century he desired.

Tipler knew that the scheme could work if, à la Van Stockum, the cylinder had an infinite length, and he also knew that such a construct was unphysical. However, he concluded that there is also a region near the surface of a *finite* cylinder where pastward time travel might be possible. His idea challenged the most basic assumptions of how the universe works, but even as a graduate student, Tipler was not one to mince words, and the paper's conclusion was unequivocal: "General relativity suggests that if we construct a sufficiently large rotating cylinder, we create a time machine."[25] In a 1976 piece,[26] he revised his conclusion and acknowledged that closed timelike curves could be generated only with a cylinder of infinite length. Since nothing infinite exists in the universe we know, this meant that this particular scheme to produce closed timelike curves was, after all, unworkable.

So ended the beginning of the first scientific investigations into time machines. To summarize, Van Stockum had imagined a scheme by which frame dragging might create a closed null curve, and he seems not to have realized that nudging the light cones a bit farther in the direction of the cylinder's rotation would make it possible to send matter and (in principle) human beings pastward. Gödel had imagined an entire universe that might operate as a time machine. Tipler modified Van Stockum's cylinder, imagining it as enabling not only signals but material bodies to travel pastward, and then reconsidered.

Finally, all these schemes assumed unphysical conditions. But there was a more fundamental reason to doubt their feasibility: each scheme described circumstances that reversed cause and effect. In the language of physics, they "violated causality."[27]

The Specter of Causality Violation

To a reader of time travel science fiction, causality violation offers opportunities for mental gymnastics, and thinking about it is, in many ways, pure and unadulterated fun. Not so for physicists. Although they have grown used to thinking about the universe as a very strange place, physicists are understandably resistant to suggestions that it is inconsistent with itself. Causality violation would mean that their most basic assumptions about nature are wrong. Berkeley physicist W. H. Williams had composed a lyric (in imitation of Lewis Carroll's "The Walrus and the Carpenter") that made witty allusions to inertial reference frames and tests of general relativity, then posed a disturbing question: if future and past are tangled,

> Then what's the use of anything;
> Of cabbages or queens?[28]

If effect did not follow cause, then all science, all logic—even reason itself—was at risk. *"What's the use of anything?"* indeed.

At least for the moment, however, those concerns were rendered academic. Although by 1976 there was no doubt that special and general relativity allowed pastward time travel, the three solutions to the field equations discussed in this chapter had been shown to be impossible in the universe we know. If three widely separated investigations into the viability of pastward time travel might be called an inquiry, then it had reached an impasse. It would be more than a decade before a physicist would publish another paper on the subject.

CHAPTER FOUR

PASTWARD TIME TRAVEL . . . SERIOUSLY

IN WHICH A SCIENTIST SEEKS ADVICE FOR A WORK OF FICTION
AND INSPIRES A NEW DIRECTION OF STUDY

Alice laughed. "There's no use trying," she said:
"one can't believe impossible things."
"I dare say you haven't had much practice," said the Queen.
"When I was your age, I always did it for half-an-hour a day.
Why, sometimes I've believed as many as six
impossible things before breakfast."
— Lewis Carroll, *Through the Looking-Glass*

E instein visited Caltech twice in the early 1930s. On both occasions he was impressed. "Here in Pasadena it is like paradise," he wrote to friends. "Always sunshine and fresh air, gardens with palm and pepper trees, and friendly people who smile at one another and ask for autographs."[1] Even thirty years later, after a population explosion and freeways, it was an agreeable environment for the Feynman Professor of Theoretical Physics, a man with a memorable name: Kip Thorne.

Thorne

In the first half of the century, physicists had used astronomical phenomena like bent starlight to test and confirm relativity, but by the 1960s the tail was beginning to wag the dog: astronomers were using relativity to explain the behavior of stars and galaxies. It was an exciting time, and Thorne counted himself lucky to be in the field. Early in his career at Caltech he explored gravity in extreme situations like the vicinity of neutron stars and the interiors of black holes, and in 1972 he formulated the "hoop conjecture," which posited that if a nonspherical imploding star with a critical mass could be completely rotated on all three axes within a hoop of a critical circumference, it would become a black hole. In 1973, with John Wheeler and Charles Misner, Thorne authored the textbook *Gravitation*. Weighing in at 1279 pages and 44 chapters, it became known as the "telephone book."[2] But its size was perhaps its only truly intimidating feature. In tone, *Gravitation* was playful and literate, and it was regarded as the classic text on gravity for many years.

Over the course of his career, Thorne has been awarded many formal honors, the most notable being his election, at the youthful age of thirty-three, to the National Academy of Sciences. But his most significant contributions are less tangible. One is an entire field of study: Thorne designed many methods now being used to detect the propagating ripples of spacetime curvature known as *gravitational waves*. The other is an international community of physicists: it is largely through Thorne's efforts that, during the cold war, Western physicists were able to collaborate with their Soviet counterparts.

Thorne is universally liked, his passion for his subject is evident, and his lectures are tours de force. Yet his modesty is legendary. When graduate students or new colleagues show undue deference, he takes efforts to dissuade them. He insists that he be called "Kip," and when he enters the room of a meeting that he is directing, his first action is often to pull his own chair away from the head of table. He had at first refused his 1991 appointment to the Feynman chair, later remarking, "It's obvious in the community that certain people are ever so much more brilliant. . . . I'm a cut or two below the Feynman level."[3] He prefers to think of himself as one of a generation of physicists. When journalists ask Thorne to list his achievements, he is likely to speak for whole minutes before they realize that he is enumerating the accomplishments of people who studied under him. Of these he is justifiably proud. Many of them, like Clifford Will and Richard Price, have made important contributions to the field.

In the 1980s, anyone meeting Thorne for the first time might have said he looked more like an anthropologist than a physicist. He wore his hair longish, often with a beard, and his taste in clothing ran to brightly colored, loose-fitting Mexican shirts. Thorne speaks softly and sometimes haltingly. Then, as now, he chooses words with deliberation, and one senses he would rather be late than be wrong.

The Question That Started It All

It was the end of spring semester, 1985. Thorne had completed his last lecture for the academic year. He was in his office in Caltech's Norman Bridge Laboratory and beginning to relax, when the telephone rang. It was Carl Sagan. Thorne had been acquainted with Sagan since the 1970s and had grown to know him better when he spent a semester at Cornell, where Sagan taught. In the 1970s and 1980s, Sagan regularly visited NASA's Jet Propulsion Laboratory in Pasadena, to be present when the first images from various unmanned interplanetary probes were received. During those visits, it became Sagan's habit to stop by Caltech to visit Thorne, and the two would grab lunch, or if there was time, Thorne would have Sagan for dinner at his home.

In 1985, Sagan was quite probably the most recognized living scientist in the world, in large part thanks to a 1973 best-selling book called *The Cosmic Connection*, and a twelve-part PBS series called *Cosmos*. His abiding interest was the possibility of extraterrestrial life and extraterrestrial intelligence, and by the late 1980s he had done more to popularize the search for extraterrestrial intelligence (SETI) than anyone before or since.

Sagan was particularly fond of speculation. He was also unflappable. Presented with the fact that no intelligible radio signals had been detected from other solar systems, he would quote Martin Rees's dictum, "Absence of evidence is not evidence of absence." When asked for estimates of the chances that extraterrestrial life existed, he would turn the question inside out and say that, given the probable number of hospitable planets, if there is no life anywhere we would be obliged to explain its *absence*. Sagan's speculations could be elaborate and detailed, and some said he took them too far. He had imagined possibilities that assumed a galaxy full of communicating societies, and he built conjecture upon conjecture, concluding long chains of reasoning with pronouncements like, "I find it difficult to believe that fewer than one hundred exchanges between the remotest parts of the galaxy would be adequate for galactic cultural homogenization."[4]

In mid July 1976, as NASA's *Viking* landers were approaching Mars, all available data suggested that the Martian environment was more like the Moon's than Earth's, and most biologists gave the chance for life there as vanishingly small. It was at Sagan's insistence that both spacecraft were equipped with cameras in case a Martian lizard (or equally exotic creature) scrambled across their fields of vision. On the question of Martian life, Sagan was what was termed an "optimist"—meaning that he thought some form of life might exist. He reserved the right to withhold judgment, and he had an interesting way of doing it. He did *not* say, "Something could live in this environment"; he said, "Nothing in these observations excludes biology." That particular syntactic construction offered a means to imply possibilities that were wildly speculative without exactly saying them. This use of the double negative became a kind of trademark, and when his critics derided him for it, he defended it as means to keep an open mind. Others suggested that Sagan's interest, and the search itself, was motivated by naïve wishful thinking inspired in turn by the likes of Edgar Rice Burroughs and Ray Bradbury. Several of Sagan's colleagues felt that, at least sometimes, he was exploiting a lowest common denominator.

Sagan was one of the few people in the long history of the National Academy of Sciences to have been provisionally elected to membership but then individually rejected in a special vote. Much of the opposition on the academy floor was framed in terms of procedural issues, or of Sagan's allegedly deficient contributions to scientific research. In fact, though, Sagan's record of publication in the professional journals was prodigious. Many believed that he lost his potential seat because he had enjoyed too much success as a popularizer. He was a charismatic presence, and remarkably telegenic. One scientist said, "It's hard for many of his colleagues to see his appearances on the Johnny Carson show or the 'Today Show' as justified by science; they feel that popularizing science is incompatible with doing science. . . . A certain amount of jealousy enters into their judgment. Heck, we'd all like to be on the Johnny Carson show."[5]

In the large view and in hindsight, Sagan's contribution to science and the public understanding of science is unrivaled. His interest in the greenhouse atmosphere of Venus contributed to our knowledge of Earth's climatic change, his work with others on Martian dust storms supplied rough models of "nuclear winter," and his articulate presentations of both subjects led to wider public awareness of human-induced global warming and the dangers of nuclear proliferation. In 1994 the National Academy of Sciences, the same organization that had denied him membership some years earlier, awarded Sagan its highest honor for "distinguished contributions in the application of science to the public welfare."

Within the SETI community, Sagan was much respected, and in some sense his passing in 1996 marked a kind of watershed, in that there is no obvious heir to the role of SETI advocate. At least there is no obvious *sole* heir. It may be that Sagan's thirst to discover life elsewhere survives, much diffused, among many scientists. Exobiology, once referred to derisively as "the science without a subject," is studied at universities worldwide.[6] In recent years, NASA has become more publicity-savvy, more willing to extol the excitement of exploration and the wonder of the cosmos, and more willing to acknowledge that the search for life is a valid motive for the exploration of space. The discovery of hundreds of extrasolar planets, riverbeds on Mars, and an ocean beneath the surface of the Jovian moon Europa are ample demonstration that the universe may be hospitable to life, and that Sagan's interests had real scientific merit.

When Thorne's phone rang in the spring of 1985, Sagan's reputation, controversy and all, was well established. He was calling Thorne to ask a favor of the intellectual sort. It seemed that the world-famous exobiologist was venturing into rather new professional territory. He was composing a novel. Not surprisingly, it would involve the possibility of extraterrestrial civilizations. But Sagan had other concerns as well. He wanted to depict the practice of science realistically, as he had grown to know it. He wanted to craft a novel of ideas in which his characters would discuss philosophy, religion, sociology, and history. Finally, he wanted to get the science right. On this count he was in need of education in the rather esoteric field of gravitational physics.

To Thorne the project seemed interesting, and Sagan was a friend. He readily agreed to review the manuscript, and it arrived in the mail some weeks later. Thorne saved it until a time when he would be mostly undisturbed, which, as it happened, was during a long car ride from Pasadena to Santa Cruz for his daughter's college graduation.[7]

The story was set in the present. Earth-based radio telescopes receive signals from an alien civilization, and embedded in those signals are instructions for constructing a sort of spacecraft. A group of scientists and engineers build the vehicle, and so provide a young SETI researcher named Eleanor Arroway a means to travel to very distant places. In the manuscript Thorne was reading, the vehicle would travel via black hole.

Thorne saw a problem immediately.

A Brief History of Black Holes

By the late eighteenth century the speed of light was known with some precision, and Newton's laws of gravitation were well understood. Thus the tools were available to make predictions regarding extreme phenomena in the heavens. On November 27, 1783, a natural philosopher named John Michell proposed to the Royal Society of London that it was possible for stars to become so compact that nothing, not even light, could travel fast enough to escape their gravity. Michell's idea was received with some interest, and the French natural philosopher Pierre-Simon Laplace posited the existence of the same phenomenon a few years later, in his 1796 work *Exposition du système du monde.*

The ideas of Michell and Laplace relied upon the "corpuscular" theory of light, but by the time Laplace's work was published there was growing evidence that light was not made of particles at all. It was thought to be wavelike, and exactly how waves of light would be affected by Newton's laws of gravity was far from clear. Consequently, the "dark stars" of Michell and Laplace were consigned to the reliquary of phenomena judged interesting but, at least for the moment, well beyond even theory. It would be more than a century before scientists would think seriously about them again.

The first to revive the idea was a German named Karl Schwarzschild. Schwarzschild was a respected astronomer who had accomplished a great deal early in his career; by age thirty-seven he had served as director of both the Göttingen and the Potsdam observatories. With the outbreak of World War I in August 1914, he volunteered for duty, and he held several posts—one in which he calculated shell trajectories, another in which he managed a meteorological station. But he did not allow such work to interrupt his own research, and even as he was serving on the Russian front he sought a solution to the field equations that would describe how spacetime might be curved in the vicinity of stars. In a matter of months he produced two papers and sent them to Einstein, who presented them on Schwarzschild's behalf to the Prussian Academy of Sciences in Berlin.

The first paper described the geometry of spacetime near a mass concentrated in a single point. Because an observer outside or on the surface of a spherical body experiences its gravity as though all its mass were concentrated at a single point at the body's center, the geometry was useful as a simplified or idealized model of a star. Curiously, only later did physicists realize that what Schwarzschild had described was a *black hole*.[8] When a body of a fixed mass shrinks in size, its surface gravity increases and the spacetime around it becomes increasingly curved. Schwarzschild posited a crucial size at which spacetime becomes so curved that it actually closes up around the body, effectively separating it from the rest of the universe. This size has come to be known as the Schwarzschild radius. Masses of all magnitudes, from galaxies to protons, have their own Schwarzschild radii, and all correspond directly to their mass. The Sun's is about three kilometers; Earth's is about a centimeter.

In the 1930s, astronomers discovered that stars were more varied than they had dared imagine, and each discovery taught them to ask new questions. One of the more intriguing was how dense it was possible for a star to become. The answer came in a series of revelations that recall the cartoon in which a floor gives away beneath a man of considerable girth, allowing him to fall through and strike another floor with a great thud, then to feel that floor give way, to fall through it and strike yet another floor, and then fall through that floor as well.

Most of the matter in our Earthly environment is made of atomic nuclei surrounded by electron clouds, and those electron clouds separate the nuclei from each other. By the 1920s, astonomers had begun to understand that stars were made of quite different stuff. A star like the Sun was composed of hydrogen and helium, and those atoms were colliding with such force that their electron shells were torn away, leaving the atomic nuclei to push directly against each other among the freed electrons. The result was termed a "plasma," a state of matter unknown on Earth. Plasma reaches temperatures in the millions of degrees, and although it has the density of lead, it behaves like a fluid.

A star like the Sun generates energy by slowly fusing the light nuclei of hydrogen to the heavier nuclei of helium, and it does so as long as it has hydrogen left to fuse. During most of its lifetime, such a star is a balance of forces: the superheated gases of its interior push outward, while the mutual gravitational forces of its parts pull inward. When the star exhausts its hydrogen fuel and cools, the outward-pushing pressure diminishes, and the inward-pushing gravitational force overwhelms it. The gravitational force becomes so strong that, although the nuclei may move as freely as ever, the movement of each of the freed electrons is limited to a "cell" with a volume thousands of times smaller than that the electron would otherwise inhabit. The outward pressure of the electrons against the walls of their cells—the "electron degeneracy pressure"—is all that prevents the star from further collapse. The star itself has become a white dwarf, a sphere about the size of Earth.

In 1930 a nineteen-year-old named Subrahmanyan Chandrasekhar, having recently been awarded an Indian government scholarship for graduate study at Cambridge University, was traveling by steamer from his native Madras, India, to England. During the passage he was not idle. Quite the contrary: he spent a good deal of his time calculating the electron degeneracy pressure for white dwarfs. Chandrasekhar had realized that the star's electrons, though traveling very small distances inside their cells, were also traveling at speeds approaching that of light, and (as predicted by special relativity) increasing their masses proportionally. This detail made a profound difference in the star's fate. The gravitational forces of white dwarfs with masses 1.4 times that of the Sun or greater would be so strong that they would overcome the electron degeneracy pressure, and the white dwarf would implode.

By the mid 1930s, Chandrasekhar had earned a PhD from Cambridge and had been named a Fellow of Trinity College. He had also greatly refined the shipboard prediction to yield an "equation of state" for white dwarfs—that is, an equation that would allow a physicist to predict how those stars would behave under various temperatures and pressures, and over a range of densities. Again, the conclusion was that stars of sufficient mass would collapse beyond the white dwarf stage.

In January 1935, Chandrasekhar described his prediction to the Royal Astronomical Society in London. By most accounts the presentation was utterly compelling. When he finished and the applause died, one man stood. It was the most imposing member of that impressive audience, the fifty-three-year-old British astrophysicist Arthur Eddington. Eddington had led the 1919 expeditions that proved general relativity; he was also author of the 1923 *The Mathematical Theory of Relativity* and the classic 1926 treatise *The Internal Constitution of Stars*. In the first half of the twentieth century, there were two leading figures in general relativity: one was Einstein himself; the other was the man who was about to respond to Chandrasekhar. For all these reasons, the audience was quiet and expectant.

Eddington began, and he spoke with a vehemence that surprised many—not the least of whom was Chandrasekhar himself. In a

matter of minutes Eddington had dismissed the younger man's prediction. He rejected the idea that white dwarfs would collapse upon
themselves, preferring to believe that a yet undiscovered "law of
nature" would intervene. In subsequent years he sought that law,
which he imagined as a process by which the stars ridded themselves of enough mass to prevent any such collapse from even
beginning. Alas, he was never to find it, and by the late 1930s
Chandrasekhar was vindicated, his limit accepted and canonized in
astronomy textbooks. For this and subsequent work, Chandrasekhar received the 1983 Nobel Prize in Physics.

In 1934, when Chandrasekhar's limit was still controversial,
theoretical physicist Fritz Zwicky and astronomer Walter Baade
published two papers describing the physics of supernova explosions, those rare but spectacular events during which a star rids
itself of so much matter and unleashes so much energy that, for a
few days, it may become brighter than the entire galaxy that contains it. Near the end of the second paper, Zwicky and Baade suggested, almost as an afterthought, that the explosion leaves behind
something far more dense than even a white dwarf: the mass of the
Sun squeezed into a sphere the size of a city. Because the sphere was
composed entirely of neutrons, a subatomic particle that had been
discovered two years earlier, Zwicky and Baade termed their hypothetical subject a *neutron star*. The prediction was disquieting to
those who resisted Chandrasekhar. The neutron star was very near
his threshold for collapse.

Meanwhile, astrophysics research in another part of the world
was proceeding under more difficult circumstances. In the Soviet
Union, Stalin's purge had by the late 1930s effected the imprison-
ment or disappearance of thousands of scientists. It was called the
Great Terror. In 1937, physicist Lev Davidovich Landau came to
believe that he might be spared if he could publish an idea that was
sufficiently groundbreaking. He did, in fact, have one such idea. He
believed that if one could strip away the outer layers of a star like the

Sun, one would reveal a core made of neutrons. Further, he believed that this core—he termed it a "neutron core"—was the source of the star's heat. With the aid of Danish physicist Niels Bohr, whose reputation as a pioneer of quantum physics afforded him considerable influence, Landau published his ideas in a 1938 issue of the British journal *Nature*.[9] Tragically, it was not enough, and in 1938 Landau was placed in a political prison, where he languished for a year.[10]

In that same year, physicist J. Robert Oppenheimer held professorships at both Caltech and the University of California, Berkeley. He read Landau's piece in *Nature* with considerable interest. Oppenheimer disagreed with Landau's idea that stars held neutron cores (and in fact, he later proved Landau wrong on that count), but it occurred to him that the neutron core was very like a neutron star. He began to wonder what happened to massive stars after they cooled, and exactly how massive a neutron star would have to become before it was overwhelmed by gravitational forces that collapsed it further.

Oppenheimer knew that gravitational forces in the vicinity of neutron stars would be extreme, that spacetime would be so warped that its behavior could be explained only through appeal to general relativity. He also realized that he needed to account for nuclear forces that held atomic nuclei together. These forces, now commonly called the "strong forces," were poorly understood in 1938. But he persevered, and with the help of physicist Richard Tolman, he formulated a fairly sophisticated equation of state for the matter in a neutron star. In 1939, Oppenheimer and Canadian physicist George Volkoff used that equation to calculate the mass at which a neutron will collapse to something even more dense. They found that mass to be somewhere between one-half and several times that of the Sun.[11]

Like Eddington, Oppenheimer would have preferred to believe that black holes did not exist. But unlike Eddington, he was willing to let himself be persuaded otherwise. He brought the Einstein field equations to bear on the problem in a particularly rigorous manner and found that they allowed no alternative. In 1939, he and his student Hartland Snyder published what is widely regarded as the first modern description of a black hole.[12]

Oppenheimer and Snyder suggested that, when an idealized spherical star implodes, even after it shrinks past the Schwarzschild radius it continues shrinking until it becomes a point with no volume but infinite density. This improbable phenomenon they termed a "spacetime singularity." There were many differences between the dark star of Michell and Laplace and the black hole as it was understood by this time, but the spacetime singularity was perhaps the most notable. A cross section of Michell and Laplace's dark star would reveal a normal star at the center of a circumference that represented its event horizon, and light corpuscles escaping the star's surface would reach as far as that circumference before falling back. A cross section of Oppenheimer and Snyder's black hole, on the other hand, would show a singularity at the center of the circumference that represented the event horizon, and between the singularity and the circumference would be only empty space.

Their black hole would have another peculiar feature. Owing to the immense curvature of spacetime inside the event horizon, any body unfortunate enough to find itself caught there and falling toward the singularity would experience a peculiar and rather spectacular demise. In the direction of its fall the body would be stretched by tidal forces to an infinite length. Physicists suspect that nothing in the real universe is infinite, and many doubted this result. The singularity itself was cause for concern for much the same reason. Although classical physicists found that their calculations sometimes yielded singularities (for instance, an electrically charged point particle has an infinite energy density in the electrical field at the point), no one thought these were real. Rather, they were considered evidence of an incomplete understanding.[13]

By the end of the 1930s research into black holes had seen a promising beginning. Had not more mundane matters intervened, it might have progressed quickly. But in the same year that Oppenheimer and Snyder's paper was published, Hitler invaded Poland, and the intellectual energies of theoretical physicists were given over to the demands of war. By 1943, Oppenheimer was in Los Alamos, New Mexico, directing the plan to design, construct, and test the first atomic bomb. No one would give serious thought to the existence of black holes again until the 1950s.

Wheeler

Perhaps encouraged by images of Einstein's disheveled hair and baggy sweater, the public has grown to expect theoretical physicists to somehow look the part. On this score John Wheeler failed miserably. Most working days he dressed in a business suit and might have been mistaken for a prosperous small-town banker. Yet his ideas, some of which we will encounter shortly, are not merely unconventional; they are among the most unusual in the history of physics. Thorne recalls Nobel Laureate Richard Feynman saying of Wheeler, "This man sounds crazy. But he *always* sounds crazy. He sounded crazy when he was my thesis advisor in the 1940s."[14] Feynman meant the description, of course, as praise of an adventurous spirit. Indeed, Wheeler's oft-quoted advice—"Find the strangest thing, and then explore it"[15]—might describe his own career.

In the 1930s, Wheeler and Niels Bohr developed a theory explaining nuclear fission. During the war, Wheeler directed the group that built the world's first nuclear reactor, and shortly after he led a design team that developed the hydrogen bomb. By the 1950s, Wheeler was a member of the physics faculty at Princeton, where he became increasingly interested in the ultimate fate of stars. He was familiar with the prewar research of Landau, Oppenheimer, Volkoff, and Snyder, and it occurred to him that the time was ripe to compile an inventory of masses and circumferences of stars that, when their nuclear fuel was exhausted, could not be reignited.

In 1958, Wheeler's student B. Kent Harrison and postdoc Masami Wakano used a digital computer called MANIAC to create such an inventory. (The unfortunate acronym stood for "*m*athe-matical *a*nalyzer, *n*umerical *i*ntegrator, *a*nd *c*omputer"; it was a machine built at the Institute for Advanced Study that had assisted in the design of the hydrogen bomb.) Anyone making even the most cursory glance at their result could draw an obvious conclusion: a star with a mass more than twice that of the Sun, unless it ejects sufficient material, will implode to something more compact than even a neutron star. To many, this finding held a rather disturbing implication. The galaxy was fairly brimming with stars that had masses more than twice that of the Sun.[16]

It was his own result, but Wheeler refused to accept it, preferring to believe that black holes were an absurdity that the laws of nature would not countenance. Like Eddington two decades earlier, he sus-pected that an unknown process removed enough mass from an imploding star to prevent such an end. Of course Wheeler knew of the results of Oppenheimer and Snyder, but he had dismissed them. The imploding star upon which they had based their calculations was idealized—perfectly spherical and of consistent density through-out. An actual imploding star is somewhat more complex. It will spin, and the resulting centrifugal force will give it a slightly oblate shape; its internal pressures and densities will vary; and as it implodes it will experience shock waves and will radiate electromagnetic waves. These last two features are the obvious mechanisms through which a real star might lose mass, and Oppenheimer and Snyder's star sim-ply did not allow for them. Nonetheless, in the early 1960s Wheeler's views began to change. Encouraged by the insights of physicist David Finkelstein and by computer simulations of implo-sions, Wheeler was beginning to believe not only that black holes were real, but that they were an unexpected gift, the natural laboratories that would enable physicists to reconcile general relativity with *quantum mechanics*—the laws of physics that explain the behavior of the universe on very small scales (the scales of molecules, atoms, and electrons) and underlie the universe on larger scales. He would later write, "Of all the entities I have encountered in my life in physics, none approaches the black hole in fascination. And none, I think, is a more important constituent of this universe we call home."[17] By 1962, with the zeal of the newly converted, Wheeler was arguing that black hole singularities might offer keys to the deepest understanding of the universe. As for the infinities of Oppenheimer and Snyder, he had a new explanation: they were artifacts of general relativity, which would be of no help in understanding what was going on inside the event horizon. Quantum mechanics, he believed, would yield more reasonable results. To understand the singularity itself, it would be necessary to reconcile Einstein's tidal gravity with quantum mechanics in accordance with another, as yet undiscovered, set of laws. These would be called *quantum gravity*.

Meanwhile, in 1961, two Soviet physicists—Isaac Markovich Khalatnikov and Evgeny Mikhailovich Lifshitz—proposed that Oppenheimer and Snyder had made an assumption that had led to a more fundamental error. The idealized density of the imploding star had allowed them to hypothesize infinite stretching of a body inside the event horizon. But the density of an actual star is uneven. Khalatnikov and Lifshitz suggested that even if differences between various densities are small, as the star implodes those variations grow, and in fact grow large enough to prevent a singularity from forming.

For several years this position was regarded as credible. Then, in 1964, British mathematician and theoretical physicist Roger Penrose realized that Khalatnikov and Lifshitz were wrong. Using methods of topology, the branch of mathematics involving surfaces left unchanged by continuous (that is, unbroken) deformation,

Penrose demonstrated that if any star, idealized or real, implodes so that its gravity becomes strong enough to pull the light rays it produces back inward toward itself, it will—indeed it *must*—collapse into a singularity.[18]

Thus, Khalatnikov and Lifshitz were provoked to rethink their work. In 1969, with a contribution from graduate student Vladimir Belinsky, they derived a solution to the field equations that offered a new and probably more realistic view inside an event horizon. Their black hole held a singularity, but of a rather different sort. A body inside the event horizon and falling toward this singularity would not be subject to a gradual and steadily increasing tidal gravity; rather, it would experience forces that were random and utterly chaotic, and it would be stretched and squeezed violently in all directions. This singularity, it should be noted, was not a mathematical point. It had actual volume, although a very small one.

Searching for Black Holes

For much of the twentieth century, the theoretical physicists called *relativists* were concerned with general relativity, and *astrophysicists* were interested in stars. Few in either group knew much of the other. Then, in 1963, when radio astronomers discovered the phenomena dubbed "quasi-stellar radio sources" (later called quasars), the knowledge of the relativists was brought to bear on the province of astrophysicists. Increasingly, astronomers were confirming predictions of theory.

In the 1960s a young British radio astronomer named Antony Hewish received a modest government grant to build a radio telescope. By 1967 it had begun operating, and in August of that year Jocelyn Bell, one of Hewish's students, noticed that its recorders had registered an unusual radio source. Astronomers knew that stars could modulate the amount of radiation they emitted, but no one had ever seen one doing it so rapidly or so precisely; this one was turning on and off every 1.33730113 seconds. Hewish and Bell realized that it was barely possible that they had detected a radio beacon from another civilization, and half jokingly (but only half) they dubbed the source LGM-1, for "Little Green Men."

When they found several sources radiating in the same manner, a natural cause seemed more and more probable. One source was in the Crab Nebula, the remnant of the supernova of AD 1054, famously described by Chinese astronomers. As it happened, this was a place that Walter Baade, some years earlier, had suggested as a likely residence of a neutron star. Indeed, when radio astronomers turned their instruments in that direction, they found a radio source at the nebula's heart, turning itself off and on at a fierce rate of almost thirty times per second. Astonishingly, the star was also radiating visible light.

In time it became clear that this star, and all the other stars in question, were not vibrating or pulsing at all. Rather, they were spinning at extreme rates, and as they spun they were sending streams of radiation out into the universe like lighthouse beacons. By the late 1960s no one doubted that neutron stars, bizarre as they seemed, were real. Black holes were less outlandish with each passing year.

Still, had the theoretical physicists thinking about black holes in the 1950s and 1960s been asked to consider an observational search, they would have responded with bemusement. Clearly there were no black holes in the vicinity of Earth or even within the Solar System, as their presence would greatly disturb the motions of other bodies. No such disturbances had been detected. A black hole alone in interstellar space would be difficult to detect because it would have no nearby bodies to affect. At least by astronomical standards, the objects in question were very small—a few kilometers wide at best. (No one had yet suggested that black holes might exist in other sizes.)[19] They were also, of course, utterly dark. Even against the background of a glowing nebula, a black hole would be difficult to see. To conduct an observational search at the distance of even the nearest star, 4.3 light-years away, would be like looking for a period on this page placed in space at a distance three times that of the Moon. For all these reasons relativists regarded such an undertaking as so dubious as to be unworthy of consideration. Then again, they were accustomed to working on blackboards, in conference rooms, and more recently with computers—not in observatories. And they did not fully appreciate the ingenuity and imagination of astrophysicists.

In the Soviet Union, Lev Landau had read Oppenheimer and Snyder's 1939 paper with great interest, and he had encouraged his colleagues to consider black holes as subjects of serious study. Among those colleagues was a physicist named Yakov Borisovich Zel'dovich. Zel'dovich was convinced that the universe was conducting experiments naturally, and if only we knew where to look, we would observe them. The big bang, he liked to say, was a poor man's particle accelerator.[20] He believed there must be a way in which a black hole would announce its presence even across interstellar distances.

Most stars are parts of binary systems—that is, two stars locked in a mutual orbit. Zel'dovich knew that the invisible member of a binary would be a candidate for a black hole, as long as it was truly invisible, and not merely one of several types of relatively massive but dim stars. Making such a determination with certainty, though, was a tall order. So Zel'dovich conceived of another sort of evidence. If the hole is traveling through a cloud of gas, then the violent collision of gas streams falling into the hole will generate X-rays. Such clouds are fairly common in the Milky Way, and in fact can be created by stars. But there were many possible red herrings: neutron stars traveling through such clouds, for instance, might by the same process also produce X-rays in great quantities. A search for X-ray sources in gas clouds seemed no more likely to identify black holes than would a search for invisible members of a binary system. Then, in 1966, Zel'dovich and his student Igor Novikov realized that they might profitably combine the two approaches. If an invisible member of a binary system was both more massive than a neutron star and also generating X-rays, it could well be a black hole.

The radio pulses of a pulsar represent an unmistakable "signature"; that is, the pulses can be explained only by the characteristics of rapidly rotating spheres of densely packed neutrons. Although physicists have imagined corresponding signatures for black holes, none so far are unequivocal. Still, most believe that they are indeed black holes, because other explanations are so elaborate and labored as to be improbable in the extreme. In the 1960s and 1970s, through the collective efforts of astrophysicists and engineers, a search was undertaken and several possibilities were identified, the most likely being an X-ray source called Cygnus X-1. At present, astronomers are reasonably certain it is a black hole, and they have catalogued several other probable black holes of masses several times that of the Sun.

When Sagan sent his manuscript to Thorne, most astrophysicists agreed that black holes did not make for a benign environment, let alone a shortcut through space. The tidal gravity of a black hole would wrench apart a traveler long before she reached the singularity, and at that place her atoms would be crushed out of existence. How, then, did Sagan come to believe that black holes might offer shortcuts through space? In small part, one of those responsible for his innocent mistake may have been Thorne himself.

In 1962, Thorne was a twenty-two-year-old graduate student at Princeton, where he had begun work under Wheeler. It was in some ways the beginning of research into black holes—what Thorne would later call a "golden age." At Princeton, as in most American physics graduate programs, students work in research groups under an advisor or mentor. Wheeler suggested to certain members of his group that they consider seriously the consequences of an imploding/exploding black hole. Among the most obvious questions was, if a star's mass explodes inside the Schwarzschild radius, where does it explode *to*? At the time it was believed that nothing below the Schwarzschild horizon could ever escape the black hole. Was it possible that the black hole somehow contained the explosion? Thorne considered a far stranger possibility: that the star explodes into another part of space, or into another universe entirely.

This notion, as heady as it is difficult, is made easier with an embedding diagram. To represent the effect of a star in space, we drop a sphere onto the membrane, and the sphere sinks into it, stretching and bending it into a sort of shallow declivity. If we now wish to represent that same star as it collapses to become a black hole, then without removing any of the star's mass, we compress it. As the star shrinks, it sinks deeper into the membrane, making the walls of the declivity lengthen and narrow until it has become a long throat with the star (now a black hole) sitting at its bottom. Finally, if we wish to represent the Schwarzschild radius, we draw a line around the circumference of the throat, just above the black hole. (Strictly speaking, the embedding diagram can represent only the region above that line, because the region of space below it, or inside the event horizon, is not static.)

In Thorne's model, a black hole could "push through" space, sever itself from it, exist momentarily as a self-contained universe, then join itself to another universe (or a distant part of the one it recently left), into which it would explode. We might imagine a ceiling and a floor, both freshly painted the same color, with a few centimeters of space between them. Paint on the ceiling collects at a certain place and gradually a drip forms, breaks free and falls, then strikes the wet paint on the floor immediately below and disappears into it.

As for the idea that a black hole could enter another universe or another region in this one, many—including Penrose—were less optimistic. In 1968, Penrose suggested that electromagnetic vacuum fluctuations and radiation fall like rain on the black hole—or at least *begin* to fall like rain—because the hole's gravity soon accelerates them to near-light velocities, and they destroy the connector, cutting off access to that other universe. But Penrose's analysis described a Reissner-Nordström black hole—that is, a black hole that is electrically charged but does not rotate. Like a pirouetting dancer pulling her arms inward, a star rotates faster and faster as it collapses. Only a very improbable set of circumstances could prevent a star from rotating, and most astrophysicists suspected that black holes rotated at a far greater rate than did the star from which they collapsed. Consequently, true Reissner-Nordström black holes were likely to be quite rare, if they existed at all, and the mechanism that Penrose believed would destroy the connector was likewise improbable.

In 1963, New Zealand mathematician Roy Kerr solved the Einstein field equations for a more realistic sort of black hole—one that rotated. Among his conclusions was the idea that the resulting singularity would not be a point; it would be a ring with zero thickness that has come to be called a "ring singularity," or "Kerr singularity." In 1968, Brandon Carter, then a research student under Dennis Sciama at the University of Cambridge, suggested that ring singularities might alter spacetime in such a way as to permit causality violation.[21] But, as many pointed out, the proof applies to the region inside the horizon, not to the region at the horizon or outside it. If such causality-violating conditions existed, they were forever hidden from direct observation and experiment.

White Holes

In the last decades of the twentieth century there developed a credible scenario for cosmic demise in which the universe undergoes complete gravitational collapse. It was termed, reasonably enough, the "big crunch"—and it had entered the public discourse. Many knew that Stephen Hawking had suggested that black holes might be thought of as small-scale versions of a universe collapsing into itself. Thus was their almost unimaginable destructive power given an apocalyptic shading. Black holes, it seemed, foretold the end of everything.[22]

Our imaginations seek symmetry and balance, and it is agreeable to believe that every death is also a birth. In the 1970s, theories of white holes and routes to other universes gave form to such notions. In the 1930s, Einstein and Nathan Rosen discovered two possible Schwarzschild solutions to the black hole, and one mirrored the other. One solution gives the black hole that in recent years has become familiar—that is, a devourer of matter and energy. The other describes it as a creator of matter and energy— a "reverse" black hole from which things may escape, but into which nothing may fall. This is a "white hole." White holes had event horizons that, like the connector between universes discussed above, were unstable against small perturbations. They would collapse upon anyone or anything trying to escape—if they were real, that is.

Einstein and Rosen's solution is thought to have been unphysical; that is, white holes seem to be impossible in the universe as we know it. How, then, can we explain the solution? Every real number has two square roots—one positive and one negative. The number nine, for instance, has as its square roots both three and minus three. But when we try to translate the second number into quantities like coconuts, we encounter difficulties. Having "minus three" coconuts makes a kind of sense to an accountant, but it makes no sense to a physicist. To a physicist the second number has no meaning, and for this reason he is likely to dismiss it. For much the same reason, most physicists dismiss the second Schwarzschild solution.[23]

Black Holes in the Popular Imagination

Most of these developments were reported in journals like *Physical Review D* and *Physics Letters B*—journals that only theoretical physicists and astrophysicists were likely to read attentively. For most everyone else, knowledge of black holes was likely to come through a generalized journal like *Nature* or a long science article in a newspaper. The latter was prone to report rather breathlessly on the first blush of a wild-sounding theory, to gloss over any reservations from the larger community of scientists, and to save any description of methodology for the last paragraphs. The same newspaper was unlikely to report on subsequent findings that undermined that theory. Such news pieces had a cumulative effect, and by the 1980s the idea of the black hole had considerable hold on the imagination of the lay public.

Black holes were repositories of nature's deepest mysteries, spacetime *in extremis*. They were also stages for dramas with Wagnerian scope, and touchstones for death and rebirth. Physicists themselves were not immune to such characterizations. In 1972, Chandrasekhar, then in his sixties, delivered the Halley Lecture at Oxford. His interests since the 1930s had led him away from black holes, but he was about to return, to develop a deeper, more complete mathematical theory. In the lecture, he alluded to the mysteries cloaked by the event horizon by reciting a story equally suited to describing the mysteries of the afterlife—the "parable of the dragonflies":

> A constant source of mystery for these larvae is what happens to them, when on reaching the stage of the chrysalis, they pass through the surface of the pond never to return. And each larva, as it approaches the chrysalis stage and feels compelled to rise to the surface of the pond, promises to return and tell those that remain behind what really happens, and confirm or deny a rumour attributed to a frog that when a larva emerges on the

other side of their world it becomes a marvelous creature with a long slender body and iridescent wings. But on emerging from the surface of the pond as a fully-formed dragonfly, it is unable to penetrate the surface no matter how much it tries and how long it hovers. And the history books of the larvae do not record any instance of one of them returning to tell them what happens to it when it crosses the dome of their world.[24]

In 1985, when Sagan asked Thorne to review his manuscript, the wilder theories associated with black holes—white holes, paths to other regions of space—made a sort of background noise in the culture to which no one could be completely deaf. Sagan was famously critical of bad science and pseudoscience; he had spent considerable energies debunking numerology, theories of ancient astronauts and flying saucers, and various claims of paranormal phenomena. Still, he was an exobiologist and a generalist. He was not an astrophysicist, and he was unlikely to peruse issues of *Physical Review D* on a regular basis. It is easy to appreciate how, like the rest of us nonastrophysicists, Sagan would have a vague awareness of such theories without knowing which were controversial or discredited.

At any rate, Thorne knew that if Sagan wanted a credible interstellar transportation system, he could not use black holes. But Thorne had been tasked with an intriguing problem, and on the return trip from his daughter's graduation, he thought about it further. By his own recollection, he was "on Interstate 5 somewhere west of Fresno" when it occurred to him that although a black hole would not serve Sagan's purpose, a *wormhole* might.[25]

Wormholes

Consider again the surface of Hinton's plane world. If a two-dimensional being, confined to that surface, wishes to travel between two distant points—say from that world's "Los Angeles" to that world's "New York"—he must travel roughly four thousand kilometers, a considerable distance. But if he could somehow bend or fold the surface measured by those four thousand kilometers in such a way that Los Angeles came *in contact* with New York, and if he could somehow push through the surface, he would have created a shortcut between the cities. Imagine now that the elastic sheet that represents three-dimensional space is gently folded, and Los Angeles and New York are brought together until they nearly touch. The short remaining "vertical" distance between the cities may be said to exist in an imaginary higher dimension called *hyperspace*. (Note that hyperspace is neither actual nor theoretical; it is purely imaginary, devised to assist our visualization.) We might connect Los Angeles and New York by a kind of tunnel. This is what physicists term a *wormhole* (Figure 4.1).[26]

When Sagan put his query to Thorne, wormholes were not new ideas. Austrian physicist Ludwig Flamm realized that the Schwarzschild solution might describe a wormhole in 1916—less than a year after Einstein formalized general relativity, and some twenty-

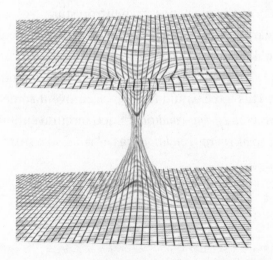

Figure 4.1 An Embedding Diagram of a Wormhole

three years before Oppenheimer and Snyder published their description of a black hole. This came to be called a Schwarzschild wormhole or, when it was further investigated in the 1930s by Einstein and Nathan Rosen, an "Einstein-Rosen bridge."

Thorne knew that wormholes might prove a better means of transportation for Sagan's heroine than black holes, but he had doubts. There were several solutions for a wormhole, and all demanded that its existence was brief. Imagine two parallel surfaces, like the folded elastic sheet just described. Each surface grows a protuberance that extends into the space between them—one downward, one upward. These protuberances touch, connect, thicken into a throat, then break apart and finally recede into the surfaces from which they emerged (Figure 4.2). The whole process occurs with fantastic speed: in fact, the wormhole throat would flicker in and out of existence so quickly that nothing—not even light, let alone Sagan's heroine—would have time to get through. Worse, like the gravity of a black hole, the gravity of a wormhole accelerates radiation, and the accelerated radiation triggers the throat to contract and seal itself shut still sooner. Before anyone could even begin a traversal, a wormhole throat would collapse with an inward pressure as great as the gravity of the most massive neutron star.

It was clear to Thorne that users of wormholes would require a means to hold their mouths open artificially with a force that would counter and slightly exceed the pressure that would otherwise collapse them. This force would have to circumvent something called the *averaged weak energy condition*. Such circumvention would not be easy. The *weak energy condition* was a "rule" in nature—it had not

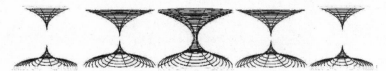

Figure 4.2 Stages in a Wormhole's Brief Lifetime

gained the status of a physical law—but it seemed to be obeyed by all forms of matter and energy. It required that, for all observers, the local energy in all spacetime locations must be greater than or equal to zero, and that no observer would measure negative energy. The *averaged* weak energy condition added a temporal dimension to the rule and thereby relaxed it a bit: the net energy must be greater than or equal to zero when time-averaged over a worldline.

Soon Thorne believed he had worked around the problem. Although the averaged weak energy condition was the basis for a number of physical theorems, it was not sacrosanct. It might be overcome with a quantity called *negative energy*, sometimes termed *exotic matter*. In our day-to-day lives, we are familiar with positive energy or mass, which generates an attractive gravitational field. Negative energy, well beyond our everyday experience (and so termed "exotic"), generates a repulsive gravitational field.

During the return trip from his daughter's graduation, on two pages of paper, Thorne scribbled out a new solution to the Einstein field equations, a solution that described a type of wormhole that, by employing negative energy, would be traversable. The answer struck him as so simple that he was surprised no one had derived it before. A few days later he wrote Sagan a long letter explaining why black holes and traditional Schwarzschild wormholes were unsuited to his needs, and offering his new wormhole design. Sagan was grateful, and he dutifully and skillfully incorporated it into the novel as part of a lively debate among relativists and astrophysicists, who decide, à la Thorne, that general relativity permits wormholes.[27]

Morris and Thorne's "Teaching Paper"

Meanwhile, it occurred to Thorne that his "traversable wormhole" solution to the field equations could be used to teach general relativity. With graduate student Mike Morris, he began putting together a paper that described his solution in greater detail. In so doing, he and Morris confronted a number of problems—and solved them.

They knew that exotic matter might be difficult to produce, so in their (three) recipes for wormholes they used it frugally, in one case limiting it to "a tiny central region . . . around the throat."[28] They suspected that the exotic matter might physically couple itself to the traveler, so they suggested several mechanisms to protect the traveler from it. They knew that tidal forces at the throat of a Schwarzschild wormhole would be as strong as those of a Schwarzschild black hole of the same mass, that unless the wormhole's mass were greater than that of ten thousand Suns, the forces acting upon a person falling toward its throat would be fatal. So they designed their wormhole such that the traveler's acceleration would never exceed one gravity and the tidal forces would be survivable. Finally, they knew that the horizon of a Schwarzschild wormhole, like that of a black hole, forever separates what is inside from what is outside, and demands that all travel across the horizon be one-way. Thus they designed their wormhole with no horizon. Of course, they were aware that they may well have overlooked some problems entirely, and in the spirit of scientific humility, they acknowledged that the wormhole might well possess additional instabilities that they had not imagined.

Designing this wormhole was one challenge; making it would be quite another. Morris and Thorne knew that the severe change in the shape of spacetime necessary would require a spacetime singularity, and that physicists may not be able to produce one without a greater understanding of quantum gravity. But they did not imagine creating a wormhole from scratch; rather, they imagined altering an existing wormhole to suit their needs. Where might one find a wormhole? The answer, strangely enough, was anywhere and everywhere.

Quantum Foam

In 1955, Wheeler and his student Charles Misner were interested in the nature of spacetime on very small scales. This area of study had few signposts. One, placed in 1900 by Max Planck, was a quantity that, when combined with the gravitational constant and the speed of light, yielded a dimension of length. It has come to be called a *Planck length*, and it is unimaginably small.

Suppose that we have a microscope with infinitely powerful magnification. Zoom in on a period on this page until it fills the view. To measure the period's diameter with atoms, we would need one million placed end to end. Zoom in on one of those atoms until it fills the view. To measure its diameter with protons, we would need a hundred thousand placed end to end. Finally, zoom in on a proton. To measure it end to end would take one million Planck lengths. A Planck length is smaller than a proton by the same factor that an atom is smaller than a period on this page. Planck himself did not know exactly what this length represented, but because it was a product of two fundamental constants, he was certain that it was important.

As it happened, he was right. Wheeler and Misner came to realize that a Planck length represented the span shorter than which space as we know it ceases to exist.[29] From this interval they derived another constant: *Planck-Wheeler time*, the time it takes for light to travel a Planck length. A Planck-Wheeler time is 10^{-43} of a second—that is, a decimal point followed by 42 zeros and a 1. Although this is a very small number (there are 10^{17} seconds in the lifetime of the universe, but 10^{44} Planck-Wheeler times in a single second), it is a fundamental one. A Planck-Wheeler time is the shortest time interval possible. If two events are separated by less time than this, it makes no sense to say which happened first and which next.[30]

Wheeler believed that, on Planck scales, space itself would have no definite shape, no definite curvature. Any depiction of spacetime on such scales would necessarily require a whole *set* of embedding diagrams. In some of these the elastic membrane would be smooth, in some it would be curved and oddly deformed, and in others it would be a surreal landscape of towers, arches, and tunnels. Still more strangely, we could not assert which of these existed at a given moment, but only that there would be a greater or lesser probability for one to exist at that moment. Wheeler called spacetime on this scale *quantum foam* or *spacetime foam*.

Because quantum foam assumed a fantastic variety of topologies, very small wormholes might be quite common. Morris and Thorne realized that, in theory, these tiny wormholes might be exploited. "One could *imagine*," they wrote, "an exceedingly advanced civilization pulling a wormhole out of the submicroscopic, quantum mechanical, spacetime foam and enlarging it and moving its openings around the universe until it has assumed the size, shape, and location required for some specific interstellar travel project."[31]

As these ideas were coalescing, the authors were engaged, naturally enough, in other activities. Morris was applying for faculty positions at universities. Thorne, in addition to carrying out teaching duties and other research projects, was chairing the steering committee for a hundred-million-dollar project being undertaken jointly by Caltech and MIT. It was the Laser Interferometer Gravitational-Wave Observatory (LIGO), a facility designed to detect and measure gravitational waves generated by black holes. For different reasons—Thorne to speak about LIGO, Morris to make professional contacts and get a firsthand sense of the profession—both attended the 1986 Symposia on Relativistic Astrophysics.

The first such symposia, a formal meeting between relativists and astrophysicists that was inspired by the discovery of quasars, was convened in December 1963, in Dallas, Texas. The Symposia since then has been held in alternating years and convened in many cities—New York, Munich, and Jerusalem among others. But at first because it was easier to say "Texas Symposia" than "Annual Symposia on Relativistic Astrophysics," and then out of sheer habit, physicists used the shorter term, although most years it was geographically inaccurate. In the two decades since its beginning, featured topics had included gravitational radiation, X-ray binary stars, and the nature of the early universe. Whatever the theme, the Texas Symposia could be counted on to show the best of new work. It was also the place to learn who was up-and-coming, and to gain a heady sense of physics at the frontiers.

A Chance Meeting, a New Idea

In December 1986, the Texas Symposia was held in Chicago. The more novel papers that year treated superstrings and dark matter. A paper concerning the weak energy condition was presented by a young assistant professor at Central Connecticut State University named Thomas Roman.[32] He had done his graduate work on negative energy and had recently published an article that suggested means by which the condition might be violated. Roman happened to meet Morris at the conference, and between presentations they began to talk. Soon enough, Morris explained the wormhole idea.

The point of shared interest, naturally enough, was the exotic matter that would hold the mouth open. Roman had doubts about it. Although exotic matter might be manufactured in small quantities, he suspected that amounts necessary to hold open a wormhole mouth were unrealistic, and in fact might well be prohibited by an as yet undiscovered physical law. As they spoke, Roman vaguely recalled a paper that, he later realized, was "Causality and Multiply-Connected Space-Time," published in *Physical Review* by Robert Fuller and John Wheeler in 1962. It had considered how a signal sent through normal space at the speed of light might be outpaced by a signal sent through a wormhole, and it raised issues of causality. Roman recalled enough of the paper that he told Morris, "You're going to have problems with time."[33]

The Paradox of Supraluminary Signals

Let us suppose that Gassendi has a technology that allows him to send signals that travel faster than light. Let us also suppose that he has won a research grant with which he has purchased, for the use of his assistant, a second ship. A paradox might arise as follows. Gassendi and his assistant, before boarding their respective ships, devise a plan. Gassendi says that he intends to send a *supraluminary* signal to his assistant, and he will do so at 1:00 PM, unless he receives a supraluminary signal from the assistant before that time. For his part, the assistant agrees to wait for the signal and promises that he will send a return signal immediately upon receiving the initial signal.

Imagine now that the two men are moving away from each other at such speeds that put them in radically separate inertial reference frames. If the planned interchange of signals were conducted at *subluminary* speeds, no unusual circumstances would result. But a supraluminary communication would produce a rather curious situation.

Suppose Gassendi sends a signal at 1:00 PM; his assistant receives it and, according to their plan, immediately sends a return signal. That signal, being supraluminary, may arrive *before* 1:00 PM. If it does, then Gassendi would not send his signal, and the assistant would not receive it and so would not send a reply. And Gassendi, because he has not received a signal from his assistant, would send his signal at 1:00 PM. It seems, then, that Gassendi will send a signal at 1:00 PM only if he does *not* send a signal at 1:00 PM.

Fuller and Wheeler's 1962 paper had concluded that the wormhole throat "pinched off," trapping the signal within it in a region of infinite curvature. As far as they were concerned, this was a satisfying and most reassuring result. The alternative would have meant a violation of causality. Between talks at the 1986 conference, Roman did not recall the paper in detail, yet he knew that if—contra Fuller and Wheeler—the wormhole could be held open, there would be rather profound consequences.

After the conference, letters were exchanged on the matter, and the question was sharpened. Back in Pasadena, Morris and Thorne added a few paragraphs near the end of their teaching paper that concluded, "Thus it would seem that if advanced civilizations can build multiple wormhole spacetimes with adjustable relative velocities, then such civilizations can use them for backward time travel and causality violation."[34] In March 1987, Morris and Thorne submitted the paper to the *American Journal of Physics* and went on to other work. The paper's take on time travel was tentative and appeared as something of an afterthought. But Thorne knew that it provoked more questions than it had answered, and the questions dogged him.

For Thorne, spring 1987 had been crammed with teaching, research, and the rather different mental effort demanded by planning and selling LIGO. He was desperately in need of time to think. Fortunately, for the spring semester of 1988 he would be on leave from Caltech and would have available long periods of uninterrupted time. His wife, Carolee Winstein, was completing a postdoc in kinesiology at the University of Wisconsin, and Thorne worked in the attic of a house they rented in Madison. He had begun a new paper that would squarely address the issues raised by the teaching paper, and he had enlisted the aid of two graduate students: Morris again, and Ulvi Yurtsever—both of whom remained in Pasadena. They communicated by computer hookup and telephone.

The Behavior of Time in a Wormhole

Of the questions that the teaching paper raised, the most compelling was how to explain the behavior of time in a wormhole. It was clear that the wormhole's inside might easily exist in a different reference frame than did the universe outside. Thorne realized that the mouths of the wormhole might remain fixed in relation to each other and so remain in the same frame, yet they might move relative to each other as seen from the outside universe and so exist in different frames.

Anything that moves is subject to the time dilation effect of special relativity, and a wormhole mouth would be no exception. If Thorne wanted a time machine that allowed him to travel ten years into the past, he could accelerate one mouth of the wormhole at near-light speeds for a period of a little more than ten years. After ten years and, say, one month, he could return the traveled mouth and place it near the fixed mouth. Like Gassendi and his assistant in Chapter 2, the traveled mouth of the wormhole would have remained nearly the same age, while the fixed mouth would have aged ten years. The wormhole, then, would have become a time machine.

Thorne could enter the fixed mouth, walk a few meters, and emerge from the traveled mouth ten years in the future. As long as the wormhole existed, he could continue using it to travel farther into the future—stepping into the fixed mouth, emerging from the traveled mouth ten years in the future, ducking back into the fixed mouth, emerging from the traveled mouth ten years farther into the future, thus skipping futureward by ten-year intervals.

The wormhole would also offer travel in the other direction. If he entered the traveled mouth, Thorne could walk a few meters and emerge from the fixed mouth ten years in the past. But in this direction there was a limit: he could go back only as far as the moment the time machine was created. Farther than that there would be no mouth to emerge *from*.

Altogether it seemed like science fiction, yet as far as Thorne was concerned, the results were clear and unequivocal. Special and general relativity allowed for the possibility of time machines. Perhaps more significantly, the wormhole time machine was different from earlier types—the Van Stockum cylinder, Gödel's rotating universe, and the Tipler cylinder (as well as Carter's ring singularity). Unlike them, it might be possible in the universe we know.

Theoretical physicists had long been accustomed to conducting thought experiments in the manner of Einstein. Indeed, in the realms they explored—the surfaces of neutron stars, the universe in the first seconds of the big bang, the interiors of black holes—they had little choice. Still, they worked within a framework that was inherently conservative, and limited to solving two types of questions. The first was, What phenomena might exist in nature as we understand it? For instance, could black holes exist? could neutron stars exist? and so on. The second type was, What experiments might physicists conduct with existing technology? For instance, might we build a particle accelerator that would accelerate electrons to a certain fraction of the speed of light?

Speculation upon the creation and maintenance of wormholes was a thought experiment somewhat more audacious than either. Thorne and his colleagues have come to call it a "Sagan-type" question. Like an Einstein thought experiment, the question asks what the laws of physics prohibit—or, put in more positive terms, exactly what is possible. But it pushes much harder against known laws. To do this it employs a hypothetical technology developed by a likewise hypothetical "sufficiently advanced civilization"—one whose technology, in a phrase made famous by science fiction author Arthur C. Clarke, is "indistinguishable from magic."[35]

This civilization comes in two forms. Either it is extraterrestrial, far older and far more technologically advanced than our own and exists at present; or it is a civilization that we humans will develop in a distant future.

The suppositions underlying these constructs were typical of Sagan's intellectual temperament. The first—the presumed existence of extraterrestrial civilizations—showed a willed playfulness.

The second—the presumed survival of humanity into a distant future—showed an inherent optimism, an Enlightenment-inspired belief that, if given half a chance, civilization and rational thought would prosper. When Thorne imagined engineering a closed time-like curve, he realized that there was something exhilarating and liberating in "Sagan-type" questions. As he was to discover, though, others were rather more wary on the matter.

Professional Concerns

In mid summer of 1987, nearly a year before the teaching paper appeared in print, Thorne's wife, Carolee, told him that she had received a phone call from Richard Price. Price had done his graduate work under Thorne and had since joined the physics faculty at the University of Utah. He was among the many of Thorne's students who had become close friends with him; at the time, they were coauthoring a piece on the "membrane paradigm" for black holes.[36] But Price had called regarding another matter. He had heard about the time machine ideas and was concerned that Thorne, to put it delicately, was losing touch with empirical reality.

Price posed his question in a jocular way, but his voice betrayed a hint of real concern. Carolee, in all good humor, reassured him that her husband was quite well and thought no more of it. But when Thorne learned of the call, he was disturbed. It was Price who had proved a few years earlier that a black hole radiates away its "hair"; that is, a deformed hole will lose its deformation, and a magnetic hole will lose its magnetic field. The man was no stranger to unconventional ideas. If even *he* regarded an interest in time machines as a symptom of mental disturbance, then perhaps the larger profession was not ready. Thorne was not about to surrender his new interest, but he resolved that his work on the subject would be especially methodical, that he and his students would anticipate and address every objection they could imagine, and that its presentation, whenever and wherever it occurred, would be impeccable.

In the spring of 1988, with the new article on time machines nearing completion, Thorne feared the worst sort of sensational and distorted reporting, and he asked the Caltech Public Relations Office to quell any publicity on the subject.[37] Meanwhile, he approached Morris and Yurtsever with an awkward issue. Both were beginning their careers, and including "time machines" in the title might prove a professional liability. But they insisted, and Thorne relented. In September 1988, "Wormholes, Time Machines, and the Weak Energy Condition" appeared in *Physical Review Letters*.

The paper declared its unusual approach outright: "Normally theoretical physicists ask, 'What are the laws of physics?' and/or, 'What do these laws predict about the Universe?' In this Letter we ask, instead, 'What constraints do the laws of physics place on the activities of an arbitrarily advanced civilization?'"[38] Traditionally, a theorist in general relativity might consider a certain arrangement of matter and energy and ask exactly how that arrangement could be expected to warp space. Morris, Thorne, and Yurtsever reversed the steps, first conceiving of how they wished spacetime to be warped and then determining the precise arrangement of matter and energy necessary to do the job.

About this time, a science writer named Keay Davidson interviewed Thorne for a piece in the *San Francisco Examiner*. The resulting article was a faithful rendering of the physicist's thinking, but its respectful tone was undercut when someone at the paper decided that it should be accompanied by a publicity still of actor Christopher Lloyd in character as "Doc Brown," the wild-eyed, shock-maned inventor of the time machine in Robert Zemeckis's 1985 film *Back to the Future*. The effect, albeit unintentional, was to suggest that Thorne and his colleagues were the crackpot scientists of legend. Thorne was troubled by the inference.

Meanwhile, *Science News* interviewed Morris, and a one-page piece appeared in that publication in early November.[39] In the weeks following, the story was reported by the *Toronto Star*, the *New York Times*, and then, in mid December, *U.S. News and World Report*. The narrative "hook" of each of these pieces was the sensational aspect of time travel, but on balance they were as accurate as pieces so brief might reasonably be expected to be. Although each journalist glossed over the more difficult physics, to his credit each also made it clear that the subject of time machines was theoretical, and that (no doubt to the disappointment of some readers) none were being constructed in university laboratories.

The moment that a scientist sees his achievement enter the public awareness is ripe with possibility. If he works quickly, he stands a chance of capitalizing on that moment, gaining notoriety for himself, his research group, and his field in general. In the short term, this notoriety might be leveraged into grant money that can fund further work; in the long term, it might attract fresh minds to the subject. Carl Sagan, for one, had made the most of his research by bringing it into the public sphere—through magazine articles, books, and television. He delighted in speculating (and, some said, speculating wildly) in full view of his readers and viewers, effectively inviting them along for the ride. Of course, Sagan worked in a field quite unlike that of Thorne. Whereas exobiology encourages great leaps of imagination and borrows from many fields, theoretical physics tends to advance more cautiously and draws from only a few closely related areas, such as astrophysics and quantum mechanics.

No one would call Thorne timid. In fact, among the qualities he admired and emulated in Wheeler was the older man's intellectual courage. "He was," Thorne said, "perhaps the ultimate risk taker. . . . I would hope that a little of that has rubbed off on me."[40] Still, probably everyone who had worked with Thorne—colleagues and students both—would describe him as meticulous, concerned not merely that his published work was correct, but that it was articulate.[41]

This perfectionist nature is no doubt part of the reason that Thorne still felt the sting of an event that had occurred some twenty years earlier. In 1966, Thorne was finishing a postdoc at Princeton and beginning a position as a research fellow at Caltech. He had calculated pulsations of white-dwarf stars, as he was to discover, wrongly. Two years later his results briefly misled astronomers into believing that pulsars were white dwarfs, and the journal *Nature* mentioned the whole unfortunate affair in an editorial. For Thorne the experience was not merely embarrassing; it was, in his own words, "ego shattering."[42]

In 1988, although Thorne had taken care to ensure that his work with Morris and Yurtsever was correct, he could not edit those writing *about* his work. The accumulation of small errors in the pieces, and the less than careful culling from science fiction, gave him pause. After speaking with the *New York Times*, Thorne asked his besieged administrative assistant to turn away further requests for interviews with a polite statement to the effect that he did not know whether time travel was possible, that work on the subject had only begun, and that he would provide a full account to the wider public when he and his colleagues arrived at a satisfactory answer.[43]

CHAPTER FIVE

PARADOX

How wonderful that we have met with paradox.
Now we have some hope of making progress.
— Niels Bohr[1]

Morris, Thorne, and Yurtsever's paper "Wormholes, Time Machines and the Weak Energy Condition" appeared in September 1988. Although most of the press took several weeks to notice, the reaction of colleagues was rather more swift. As for Richard Price, Carolee had successfully assuaged his fears for her husband's sanity. Nonetheless, he was concerned that Thorne's new interest would tarnish an impeccable and hard-earned record of scholarship, and he said as much. Thorne waited for other reactions. Soon enough, queries began to arrive, questioning some points and asking for elaboration on others. The journal *Nature* published a brief summary of their paper by John Friedman, a physicist with whom Morris had recently begun postdoctoral work.[2] The piece was flattering and suggested that their treatment had been rigorous. Thorne was relieved: it was clear that the physics community was prepared to take them seriously.

Better still, one colleague thought the paper cause for pure cele-
bration. He was Igor Novikov, the Russian physicist who, with
Yakov Zel'dovich in 1966, had devised the ingenious means of
identifying black holes discussed in the previous chapter. Novikov
had been thinking about closed timelike curves for years, and now
Thorne's work granted him license to make such thoughts public.
In a telephone call to his friend he said, "I'm so happy, Kip! You have
broken the barrier. If *you* can publish research on time machines,
then so can I!"[3]

Novikov

Like many other physicists, Novikov possessed a sense of awe at the
universe that, to nonphysicists, seemed almost childlike. That he
maintained the outlook in the face of an early life beset with hard-
ship is itself a marvel. Novikov's father was an official in the People's Commissariat for Transportation. In 1937, during Stalin's
Great Terror he was arrested and executed. Novikov was two at the
time and remembers nothing about him. A second tragedy followed: his mother was exiled to the Gulag. When she was released
some years later, she was forbidden to visit her children; and
although she managed to see them in secret, it was always with
some fear for her own well-being. Novikov wrote that she lived out
her life in terror of being returned to prison, struggling to understand the reason for her arrests.

For reasons as unfathomable, Novikov and his elder brother were
"children of an enemy of the people"—a phrase common during
the Stalin regime, the oft-cited rationale for such a category being the
Russian proverb about the apple never rolling far from the tree. The
boys were raised by their grandmother. During World War II the family was evacuated to the village of Krasnokamsk on the Volga.

Although food was carefully rationed and luxuries were nonexistent, the grandmother somehow managed to find books. One day she

PHYSICISTS AND WAR

In the twentieth century, the work and lives of physicists were greatly shaped by developments in the political sphere, and by war. Karl Schwarzschild, the astronomer who discovered the first exact solution of Einstein's equation of general relativity, died on the Russian front while serving in World War I. Arthur Eddington, the preeminent astrophysicist of the first half of the twentieth century, was a practicing and devout Quaker and requested deferment from the First World War on that basis. The elder son of German physicist Max Planck died in battle in 1916, Planck's younger son was killed by the gestapo, and his Berlin home was bombed in 1944. Planck himself had been informed by Hitler that only his age prevented him from being sent to a concentration camp.

Einstein, of course, was rarely far removed from these matters. He produced his general theory of relativity during World War I. His private letters and journals show that science offered him both community and sanctuary, but at moments even this was not enough. In 1917 he wrote, "I cannot help being constantly terribly depressed over the immeasurably sad things which burden our lives. It no longer even helps, as it used to, to escape into one's work in physics."[a] Einstein's successors have also been affected by the political. During World War II, the Germans and the British had a mutual agreement to spare centers of learning: if the British did not send the RAF over Heidelberg and Göttingen, the Germans would not bomb Cambridge and Oxford. Stephen Hawking's parents decided that his mother would carry him to term in Oxford because it offered a relatively safe environment.[b]

If politics affected physicists, then physicists also affected politics. Indeed, Einstein's worldwide fame was in part the response of a world exhausted from war. The physicist was living proof that humanity was capable of better. In the 1920s, Einstein supported a universal disarmament; when in 1933 Hitler became the German chancellor, Einstein resigned from the Prussian Academy and resigned his German citizenship; after World War II he supported ideals of world government. At various points in his life he was an active pacifist, a spokesman for moral Zionism, and (often) a target for anti-Semitism.

In the 1940s the best American physicists were summoned to the

Manhattan Project, the effort to design and construct American bombs. As might be expected, their views of the experience differed enormously. J. Robert Oppenheimer, who led the project, described its history with an anguished eloquence: "The physicists have known sin; and this is a knowledge which they cannot lose."[c] But physicist John Archibald Wheeler, who had lost a brother in the war, expressed a rather different regret, observing that if the bomb had been developed two years earlier, millions of lives would have been saved.[d] There was, too, a way in which the wartime work assisted science. In 1964, U.S. and Soviet researchers adapted bomb design computer codes to simulate stellar implosions. They were in some ways beating plowshares from swords—swords, they could never forget, that they had made.

Physics in the Soviet Union during the twentieth century is a dramatic story in itself; the lives of Russian physicists are a catalogue of human tragedy played out against a backdrop of wars and tyranny. The early life of Kip Thorne was far more comfortable, but he was acutely aware of the suffering of his counterparts. His father was among the first American scientists to visit the Soviet Union after Stalin, and Thorne himself continued the tradition of open exchange, orchestrating meetings between Western and Soviet physicists during the cold war. The involvement of physicists in worldly matters continues. In November 2004, for example, Hawking spoke at a demonstration in London's Trafalgar Square denouncing the United States–led invasion of Iraq.

[a] Pais, *"Subtle Is the Lord—"*, 243.
[b] Hawking, *Black Holes and Baby Universes*, 1–2.
[c] "Physics in the Contemporary World," Arthur Dehon Little Memorial Lecture at the Massachusetts Institute of Technology, November 25, 1947.
[d] Wheeler, *Geons*, 18.

brought home a children's encyclopedia. When the young Novikov examined its brief section on astronomy, he found several illustrations, one of which was a full-color representation of a solar flare, with a pea-sized Earth placed alongside to provide a sense of scale. The image struck the child with a force that was almost physical.

He would later write that at that moment it became clear to him that the universe was magnificent beyond imagination.

In the summer of 1965 Novikov was thirty years old, making his first journey abroad and speaking at the Fourth International Conference on Gravitation. It might have been an anxious moment: his first presentation to the world community of physicists, delivered to a room of several hundred. By all accounts, however, he need not have felt intimidated. Thorne, then a twenty-five-year-old newly minted PhD, was in the audience and was stunned by the force and scale of Novikov's ideas.[4] When Novikov had finished, several physicists surrounded him, asking for details and elaboration. Novikov recalls that one was rather different—"a tall, lean reddish-haired youngster" who seemed American.[5] Novikov's English was far from fluent, and he was having difficulty communicating his thoughts. The "youngster"—it was Thorne—had studied Russian as an undergraduate and offered to assist.[6]

When the others left, the two continued talking and discovered that they shared a great deal, and that their approach to problems was remarkably similar. In ensuing years, despite separation by thousands of miles and governments that allowed scientific exchange only occasionally, they developed a theory explaining the structure of the stellar remains that surround black holes—phenomena known as accretion disks. Thorne edited translations of books by Novikov, and Novikov edited the Russian translation of Misner, Thorne, and Wheeler's *Gravitation*. By the late 1980s, when their conversation involved relativity and black holes, one could finish a thought that the other began.[7]

Thorne, Morris, and Yurtsever were aware of the problems that time machines presented to causality and had deliberately sidestepped them. Although their paper's conclusion acknowledged, "This wormhole spacetime may serve as a useful test bed for ideas about causality, 'free will,' and the quantum theory of measurement,"[8] they did not speculate on such matters in the text itself. In November 1988, Thorne told a *New York Times* reporter, "We're not forced to confront any philosophical considerations because this is theoretical physics, not philosophy."[9]

Nonetheless, to theoretical physicists thinking about closed time-like curves, causality was not merely an issue for philosophers. Rather, it was the elephant standing in the parlor. In fact, it had already been addressed by Novikov in his 1979 book *Evoliutsiia vselennoi* (*Evolution of the Universe*). Thorne had overseen the editing and production of its 1983 English translation, but he had not read it with the care he would have observed had he been editing it himself. Consequently, he did not recall that one of the final chapters contains the statement, "The close of time curves does not necessarily imply a violation of causality, since the events along such a closed line may be all 'self-adjusted'—they all affect one another through the closed cycle and follow one another in a self-consistent way."[10]

The Self-Consistency Principle

This was the first appearance in print of Novikov's precept that a signal or material body sent pastward would not and could not create an inconsistency. A pastward time traveler would be unable to alter history; rather, he would be shown to have "always" been part of that history. And although he would create no paradox, he would produce some very peculiar situations, some of which are illustrated by *Ascending and Descending* (Figure 5.1), a famous lithograph by Dutch artist Maurits Cornelis Escher.

Escher's image is endlessly fascinating, and his own comments on it are as evocative as we might expect: "It may be part of [the monks'] daily ritual duty to ascend the stairway in a clockwise direction during certain hours. When they are tired, they can change direction and descend for a while. But both motions, though not without an abstruse meaning, are equally useless."[11]

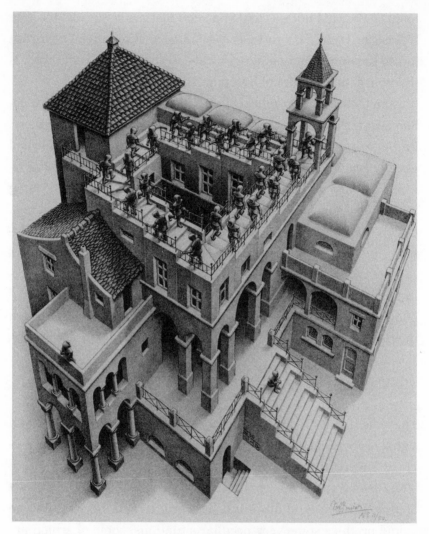

Figure 5.1 *Ascending and Descending*

Although Escher nowhere suggests such a connection, the image has characteristics of a self-consistent closed timelike curve. Each monk on the staircase is both ahead of and behind every other monk; we can identify no particular monk, and no particular stair, as "first" or "last"; and although any place in the sequence may allow an interruption, each allows it equally. Likewise, each event on a closed timelike curve is in both the past and the future of every other event on the curve; there is no "first" event or "last" event.

The inspiration for Escher's work was not closed timelike curves; it was, however, from a physicist—specifically, a young Roger Penrose. Recall that Penrose had derived his conclusion that singularities lived at the center of black holes by applying the methods of topology. The field had held his interest for years. Even as a boy he had been intrigued by shapes, especially unfamiliar ones. In one inspired moment he had imagined something he called an "impossible object"—that is, a form that could be rendered in two dimensions but that could not exist in three dimensions. He called it a "tribar" (Figure 5.2a). With the help of his father, a professor of genetics at University College in London, Penrose developed it into a second impossible object: a Penrose staircase (Figure 5.2b).

Some years later, father and son coauthored an article that described both objects. It is an especially interesting piece, as we might expect the collaboration of a physicist and a geneticist appearing in the *British Journal of Psychology* to be.[12] It acknowledges Escher, whose art depicted structures and landscapes that were similarly "impossible."

That the younger Penrose knew and admired Escher's work is not surprising, because by the mid 1950s Escher prints had grown increasingly popular, especially among mathematicians. It is per-

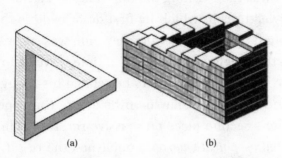

Figure 5.2 The Tribar (a) and the Penrose Staircase (b)

haps more remarkable that Escher learned of Penrose. In 1959, some of Escher's friends sent him a copy of the Penrose article. He was flattered at the allusions to his work and intrigued by the two objects that the Penroses described. A year later he produced the more elaborate version of the Penrose staircase—his *Ascending and Descending*—embellished with ambulatory monks.

The Grandfather Paradox

Novikov was not the first to suggest that, if pastward time travel were possible, it would be self-consistent. In fact, several philosophers had arrived at a similar conclusion. The best known of these was Princeton's David Lewis. Lewis had long been admired for his originality, the care with which he explained his thinking, and perhaps more than anything, his intellectual range. He published important work in ethics, political philosophy, the philosophy of science, and—this the broad category under which time travel might fall—metaphysics. But Lewis nourished interests well outside even these.

He kept a detailed scale model of the British Rail in his basement, and when he visited England he spent his days riding trains. In the late 1960s he developed a love for Australia and all things antipodal, including Australian-rules football. With his wife, Stephanie, he made annual visits to that country, and in time they came to consider Melbourne a second home. It was during a 1971 visit that Lewis, then visiting the University of Adelaide, delivered a series of lectures in which he first described his thoughts on travel through time. No one who knew him was surprised that he compared the looping worldline of a time traveler to a curving train track. In the ensuing years he refined his thoughts on the subject, and in 1976 (a few months before the appearance of Frank Tipler's second piece on massive rotating cylinders), the *American Philosophical Quarterly* published the result: an article entitled "The Paradoxes of Time Travel."

Chapter 3 illustrated the most well-known of these paradoxes using Kurt Gödel and his younger self. Let us here employ the more familiar, indeed the "classic," case—so familiar that its varieties have been explored in hundreds of science fiction stories.[13] Simply put, if I travel back in time before my birth and kill my grandfather, I cannot have been born. If I was never born, then I cannot have traveled back in time to kill my grandfather. But I had to kill my grandfather in order to bring about the circumstances in which I was never born. And so forth.

This paradox has many permutations, but all require only that I prevent my birth. Most versions are accomplished by means less dramatic and with considerably less violence. I may, for instance, merely take action to prevent my grandparents from meeting. Or I may prevent my parents from meeting. The advantage of using grandparents over parents, and homicide over its less violent alternatives, is that they offer a greater degree of certitude. The simplest and most straightforward permutation contracts the time to a few minutes and restricts the cast of characters to one: I travel five minutes into the past and kill myself. The same paradox unfolds.[14]

Lewis's article trods this familiar ground again, asking us to consider the case of "Tim," whose grandfather made a fortune in the munitions trade and whom Tim, a pacifist, detests. Tim trains himself in marksmanship, and with rifle and ammunition travels back through time to an opportune date in his grandfather's lifetime. He makes careful study of the route his grandfather takes to the munitions factory each day. He secures a room in a building along that route and stations himself at a window that allows a clear shot. Then he sets his rifle on the window ledge, looks through the rifle's sight, and waits. At precisely the moment expected, he sees his grandfather's form enter the sight's crosshairs. Here, of course, the paradox presents itself. If Tim does kill his grandfather, then Tim cannot have been born and so cannot live to kill his grandfather, in which case his grandfather lives, and so on.

Lewis presents his interpretation of the problem by stages, in the careful manner of a logician. He admits that there is a set of conditions that make it possible for Tim to kill his grandfather: Tim's marksmanship is good, his rifle is clean and well lubricated, the door behind him is locked and bolted against intruders, and his grandfather keeps to his route.

However, Lewis says, there is also another, larger and more inclusive set of conditions that make it impossible for Tim to kill his grandfather; among these are his grandfather's lifetime subsequent to the moment he enters the sight's crosshairs and, of course, Tim's very existence. In fact, says Lewis, the whole history that Tim is trying to "unmake" is itself proof that he will fail. So exactly *how* does Tim fail? Lewis suggests numerous possibilities: "Perhaps some noise distracts him at the last moment, perhaps he misses despite all his target practice, perhaps his nerve fails, perhaps he feels a pang of unaccustomed mercy."[15] The particular "how" is unimportant. The point is that he must fail, or to put it most succinctly, he *will* fail because he *did* fail.

The article appeared some five years before Novikov's first discussion of self-consistency. There were differences. Lewis treated temporal paradoxes as a problem in logic; he did not address the challenges to physics. There was a matter of differing nomenclature: what Novikov calls "closed timelike curves" Lewis had called "causal loops." Yet at its core, Lewis's position—that pastward time travel might be self-consistent and thus present no paradox as such—was fundamentally identical to Novikov's.

But the idea of self-consistent loops through time did not begin with Lewis either.

Time Travel Science Fiction and Self-Consistency

The corpus of time travel science fiction includes scores of stories in which a time traveler visits a past intending to make a change and finds that, much as he tries, something stops him. Attempting to prevent Lincoln's assassination, for example, he slips, a door is locked, or someone moves in his way. Even when his actions have some effect, they do not create a detectable change in the history he had known, but instead become woven seamlessly into that history. Much of the pleasure of reading such stories derives from the ingenious means by which this weaving is accomplished.

In one of the more famous, a paleontologist and his assistants begin to uncover a trail of fifty-million-year-old footprints made by a large predator. From the footprints' wide spacing, they judge that the animal was chasing prey. Physicists nearby are experimenting with time travel, and with their help the paleontologist takes an expedition jeep back to the time and place of the chase. His assistants return to finish the work and are wondering where he has gone—until they reach the end of the tracks to discover, beneath fifty million years of sediment, a distinctive tire tread half obliterated by a dinosaur footprint.[16]

In many of these stories, meddling in the past influences a well-known history, or actually brings that history about. A device sent back in time to record the paranoid behavior of historical figures provokes that behavior. An infrared camera intended to record texts lost in the fire that destroyed the Royal Library of Alexandria ignites that fire. A time traveler obsessed with the Crucifixion manages to travel to its place and time and is mistaken for Christ, or in a manner of speaking *becomes* Christ, and is himself crucified, thus setting into motion a religion and (not incidentally) his own obsession. Stories of self-consistent time travel take on a variety of tones. In one of the more whimsical, a movie director arranges to go "on location" to film the Vikings discovering America and finds that they do not appear when expected. Facing a deadline and losing money, he imports extras from his own time. When filming is completed, the "real" Vikings are still nowhere to be seen, the extras elect to stay, keeping their costumes and props, and history as we know it ensues.[17]

By the 1950s there were already hundreds of such stories in print, and so many permutations of the plot device that it had generated its own parodies. In 1959 a new animated television cartoon series featured an articulate, professorial dog and his "pet boy" who used a time machine to visit various figures in history.[18] In a typical episode the historical figure is discovered to be a sort of well-meaning but hopelessly misguided bungler; only with astute guidance from boy and dog (more the latter than the former) does he accomplish the deeds for which he becomes known. Thus the "great man" theory of history is quietly subverted, and history as we know it is preserved.

Lewis knew something of time travel in science fiction, and he acknowledged that, despite the genre's uneven reputation, some writers had approached the subject with rigorous logic. His 1976 article cites two stories in particular, both by the well-known (and sometimes controversial) author Robert A. Heinlein.

In the first, the 1941 "By His Bootstraps," a doctoral student working on a thesis disproving time travel is accosted by two men who are future versions of himself and, not incidentally, tangible evidence that his thesis is deeply flawed. As the story progresses, there is much jumping between times and many more versions of the student. Collectively, the versions use knowledge gained in time travel to amass political and military power until, some thirty thousand years hence, one of them reigns as a kind of global dictator. The reader then realizes that the title of the story, an oft-used description of that characteristically American career, has become darkly ironic.

What is most interesting about the story, however, is not its ending, but how it gets us there. A given scene is played once, then replayed from another point of view with added detail, then replayed again from yet another point of view with still more detail. As with a fugue, we hear a motif established, then reintroduced with some elaboration, and then reintroduced with so much elaboration that we need a moment to recognize it.

Not all self-consistent time travel stories operate on such grand scales. Some loops are almost entirely personal, but no less intriguing. One of these happens to be the other story to which Lewis's article pays homage: Heinlein's 1959 "—All You Zombies—." The author himself called it the "Furthest South" of temporal paradox stories,[19] a claim that would be difficult to dispute. The narrator is an agent for the "Temporal Bureau." He journeys forward and backward in time, meets versions of himself at various ages and (because he underwent a sex change operation) in both genders. He seduces himself, and becomes his own mother, father, daughter, and son.

In the hands of a less capable author such a premise might descend into mere gimmickry, yet thanks to particulars and tone it does not. Heinlein paints a world that is disturbingly noirish and so convincing in detail that it seems lived in. Here the narrator describes the time machine:

> I opened a case, the only thing in the room; it was a U.S.F.F. Coordinates Transfer Field Kit, series 1992, Mod. II—a beauty, no moving parts, weight twenty-three kilos fully charged, and shaped to pass as a suitcase. I had adjusted it precisely earlier that day; all I had to do was shake out the metal net which limits the transformation field.[20]

Indeed, for the narrator his world is too familiar; he has worked as a temporal agent so long that he can no longer find inspiration in the Temporal Bureau's bylaws. For the reader, though, those bylaws induce a mild and not unpleasant disorientation, including as they do, such neatly inverted maxims as "Never Do Yesterday What Should Be Done Tomorrow," and "It's Earlier When You Think." They also include reminders of the subtle power an agent wields, like "A Paradox May be Paradoctored." Such narrative finesse was of some interest to Lewis. But what most impressed him, and many readers since, is that for all the complexity and involution of the story's causal loops, they are utterly logical and completely self-consistent.[21]

It is probably no coincidence that stories of time travel first appeared in significant numbers in the late nineteenth century, when most of the globe had been mapped and there seemed few places left to discover. Futureward time travel promises a new frontier. Pastward travel—to pasts that could be, as it were, discovered again—promises another. But pastward travel has additional enticements, and these may be the reasons that, of the two directions of time travel, it is more common in science fiction. It appeals to the historian (like Wells's "Psychologist") who wishes to gain knowledge of an event directly, without intermediaries; to the autobiographer who wishes to revisit a moment and compare it with his memory of it; and to the merely nostalgic, who longs for a return to an innocent and unfallen world and a restitution of what is lost or believed lost. Of course, pastward time travel might offer more than an opportunity to merely witness. It might grant a second chance to set things right and to save whatever (or whomever) was lost. The same opportunities, much dramatized in science fiction, appeal to baser instincts as well. For instance, a time traveler might change the past in ways that give him otherwise unattainable wealth and power.

Most stories of pastward time travel provide the pleasure of a satisfying plot twist. But their real appeal, at least so it seems to me, is more enduring. They lead us to rather deeper inquiries, to questions of the reality of fate, and to the mystery of free will.

Pastward Time Travel and Free Will

Philosophers, theologians, and indeed most of us regard free will as necessary to a meaningful existence. Yet its status has seldom been secure. In the nineteenth century, the concept of free will was at odds with *determinism*, a doctrine that, proceeding logically from Newton's mechanistic universe, claims that every event is the result of previous conditions, and that all events are, in theory, predictable. Its clearest and strongest formulation was from Laplace's 1814 *Essai philosophique sur les probabilités* (*Philosophical Essay on Probabilities*), which asserted that if a superhuman intelligence knew the state of the entire universe and all its parts at a given instant, that intelligence would be able to predict all past and future states of the universe as a whole, as well as states of any of those parts.

Free will was dealt a further blow in the twentieth century when Einstein's relativity suggested that ideas of a present or "now" are local and illusory, and that worldlines do not grow into being but rather exist always in their entirety. This model of the cosmos is sometimes called a *block universe* because past, present, and future are imagined to be held within it, as though embedded in a single, four-dimensional block. Its implications, at least to some, are disturbing. If all worldlines are fixed in spacetime and the present is an illusion, then I should expect to have no more free will in the future than I do (or did) in the past. The difference—the *only* difference— is that because I do not know the future, I cannot know exactly how my free will is restricted. Nonetheless, in the block universe I trust that it is restricted, and that what will happen will happen because it cannot be otherwise.

Physicists of the twentieth century were fully aware that relativity and free will, at least as they are traditionally understood, cannot coexist. Eddington arrived at the conclusion as though backed against a wall. He wrote, "Like most other people, I suppose, I think it incredible that the wider scheme of nature which includes life and consciousness can be completely predetermined; yet I have not been able to form a satisfactory conception of any kind of law or causal sequence which shall be other than deterministic. It seems contrary to our feeling of the dignity of the mind to suppose that it merely registers a dictated sequence of thoughts and emotions; but it seems equally contrary to its dignity to put it at the mercy of impulses with no causal antecedents."[22]

It is interesting that Einstein considered these conceptions small-minded and petty, and believed that existence in the block universe, where all was determined, was far more free than the alternative.

> The man who is thoroughly convinced of the universal opera-
> tion of the law of causation cannot for a moment entertain the
> idea of a being who interferes in the course of events. . . . He has
> no use for the religion of fear and equally little for social or moral
> religion. A God who rewards and punishes is inconceivable to
> him for the simple reason that a man's actions are determined by
> necessity, external and internal, so that in God's eyes he cannot
> be responsible, any more than an inanimate object is responsible
> for the motions it undergoes. . . . A man's ethical behavior should
> be based effectively on sympathy, education, and social relation-
> ships; no religious basis is necessary. Man would indeed be in a
> poor way if he had to be restrained by fear of punishment and
> hope of reward after death.[23]

In any case, it is difficult to imagine how the block universe can exist alongside free will. It seems that if we accept the universe Einstein gave us, we must surrender free will. If we insist upon having free will, we must surrender Einstein's universe.

The issue is not as removed from the everyday as one might think. Systems of reward and punishment assume the reality of free will, as do many religions. But the reason the issue never elicits more than a passing interest from any but philosophers is that, from any particular point in spacetime (and we all exist at a partic-ular point in spacetime), the future is unknown and unknowable. Thus the *possibility* of free will is preserved and has allowed many of us to remain comfortably agnostic on the matter.

The prospect of pastward time travel, though, throws the ques-tion into sharper focus. Time travel might allow us actually to *test* the concept of free will, at least under some circumstances, and at least as a thought experiment.

In Einstein's spacetime, the outcome of that experiment should be obvious: a time traveler visiting the past would find his free will severely restricted. He would be unable to change the past. Some have suggested that this constraint on free will is in itself reason to doubt the feasibility of time travel. In their 1973 work, *The Large Scale Structure of Space-Time*, Stephen Hawking and George Ellis wrote, "The existence of [closed timelike curves] would seem to lead to the possibility of logical paradoxes. . . . Of course, there is a contradiction only if one assumes a simple notion of free will; but this is not something which can be dropped lightly since the whole of our philosophy of science is based on the assumption that one is free to perform any experiment."[24]

Novikov's rather straightforward reaction to the concern was to dismiss it as naïve. He noted that gravity forbids us from levitating or walking on ceilings "without special equipment," but few have bothered to declare it an enemy to free will.[25] All laws of physics, he pointed out, severely restrict free will, but we are so accustomed to them that it does not occur to us to say as much. It was obvious, Novikov said, that a person who traveled to the past would be unable to perform certain actions. But this was hardly a reason to think pastward time travel impossible.

In 1988, Thorne and his students had deliberately avoided using a person in the closed timelike curve, thus leaving the issues of free will for others. As to causality, they had searched for an irresolvable paradox and found none. It seemed that pastward time travel, at least for anything without free will, did not violate logic—or so they thought.

Then, some weeks after their paper's publication, the authors received a letter from physicist Joseph Polchinski at the University of Texas in Austin. If the normal evolution of a theoretical model was from idealized to complex, then Polchinski had neatly reversed it. He had taken the grandfather paradox, removed from it the metaphysics and complicating issues of free will, and distilled it to essentials—that is, to a simple problem in nonlinear algebra. Polchinski made its subject what a later paper would call "a single, classical particle that carries a hard-sphere, repulsive potential and has no internal degrees of freedom."[26] In lay terms, it was an everyday object that would obey Newton's laws of motion—a billiard ball.

Polchinski's Paradox

Imagine a wormhole time machine of the type first suggested by Thorne and Morris that is relatively modest in scale. It is the size of a billiard table, and in place of two adjacent corner pockets are the mouths of the time machine/wormhole. The left mouth has aged normally along with the outside universe, while the right mouth has been subjected to one second's worth of time dilation. Therefore, as measured from the outside universe, anything entering the right mouth would emerge from the left mouth exactly one second in the past.

There are, of course, many possible trajectories that the billiard ball might take into the right mouth and out of the left, and one second into the past. In his letter, Polchinski argued that among those trajectories must be one in which the billiard ball emerging from the left mouth struck its earlier self *before* it entered the right mouth, deflecting it from its course and thereby preventing it from entering the right mouth to begin with. Of course, if the ball did not enter the right mouth to begin with, it could not have struck its earlier self and deflected it from its course. And if it did not strike its earlier self and deflect it from its course, that earlier self could not have entered the right mouth, and so on.

Polchinski called this the "paradoxical" trajectory, and Thorne and his students began to refer to the whole scenario as "Polchinski's paradox." It was clear that it violated any presupposition that cause must precede effect, insofar as the "effect" of the struck billiard ball was brought about by the entry of the billiard ball into the right mouth—an event that occurred in its own future. Equally problematic, it represented a phenomenon that was not consistent with itself. The ball would have not one, but two worldlines—one that traced a loop, another that traced no loop. If the loop worldline existed, the nonloop worldline could not; if the nonloop worldline existed, the loop worldline could not.

Thorne, Morris, and Yurtsever began to think seriously about Polchinski's billiard ball, and the possibility that there existed one trajectory with a self-consistent solution. By late 1989, though, Morris and Yurtsever had completed their PhD's. Morris left for a postdoc at the University of Wisconsin in Milwaukee, and Yurtsever for a postdoc at the International Centre for Theoretical Physics in Trieste, Italy. Thorne, meanwhile, was not prepared to abandon Polchinski's paradox, and he presented the problem to two new students in his research group—a Chilean named Fernando Echeverria and a German named Gunnar Klinkhammer—who attacked it with enthusiasm.

Echeverria and Klinkhammer determined early on that their work would not require an actual pool table; indeed, such would have complicated the problem by forcing them to account for imperfections in the balls or table surface, as well as changes in temperature and humidity across the path the ball would travel. Instead, they modeled the trajectories with linear algebra. They were fairly confident that they would find a self-consistent solution. Soon enough they did. It was this: The billiard ball begins traveling toward the right mouth along the same trajectory traveled by Polchinski's ball. Before it reaches the mouth, it is struck by a second ball with a gentle glancing blow to its left rear side. It is thereby given a counterclockwise spin, and its trajectory is changed slightly—not enough to make it miss the mouth, but enough that it enters at a slightly different angle in a slightly different place. This

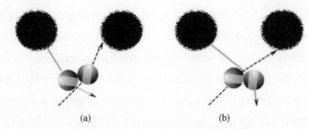

(a)   (b)

Figure 5.3 Two Solutions for a Billiard Ball and a Closed Timelike Curve

change is such that the ball emerges from the left mouth (it is now the "second ball" mentioned earlier) at an angle slightly different from that of Polchinski's ball. It does not strike its younger self directly, as did Polchinski's ball. Rather, it strikes a gentle glancing blow on its left rear side. (This solution appears as Figure 5.3a.)

An Unexpected Finding

Recall that classical Newtonian physics teaches that, in principle at least, every experiment is replicable. If we set up the same conditions and put the same experimental apparatus in motion, we will produce the same results. The only reason billiard shots in the actual world are not easily replicable is because of uncontrollable variables mentioned above. If we could suppress those variables— and in the idealized world of linear algebra we can—then if we send one billiard ball toward another on a given trajectory and at a given speed several times, each time it will strike the other in exactly the same way, and the other ball will react in exactly the same way. That result will be the same no matter how many times we repeat the shot. Thus, Echeverria and Klinkhammer were not surprised to discover that a particular shot would produce a particular solution. They *were* surprised, though (and Thorne was surprised too), that the same shot produced *another* solution.

In the second solution (Figure 5.3b), the billiard ball begins by traveling toward the right mouth along the same trajectory as the ball in the first solution, and before it reaches the mouth it is struck by the second ball. This time, though, the second ball begins to pass *in front of* the first ball but does not quite clear it, so the first, younger ball strikes it with its right forward side. As in the first solution, the younger ball is given a "spin," but it is clockwise. This time its trajectory is changed by a small angle—not enough to make the ball miss the mouth, but enough to make it enter at a slightly different angle in a slightly different place.

It was clear that the two solutions were a result made possible only by the presence of a closed timelike curve. The result was deeply troubling because it suggested that time machines make things behave in ways that seemed impossible. If there were any suspicions that the two-solution result was an artifact of the particular experiment and not characteristic of all closed timelike curves, they vanished when, in a meditative moment, Thorne imagined another two-solution trajectory for a billiard ball.

In this scenario, as before, two wormhole mouths are positioned side by side, and the right mouth has experienced one second's worth of time dilation, such that anything entering it will emerge from the left mouth one second in the past as measured from the outside universe.

In one solution (Figure 5.4a), the ball follows a trajectory between the mouths and simply continues traveling along its trajectory, untouched. In the other solution (Figure 5.4b), at the moment the ball reaches a point between the mouths, it is struck on the left side by a ball that emerged from the left mouth. The impact sends the ball into the right mouth, and it emerges from the left mouth one second earlier, in time to knock its younger self into the right mouth and be deflected along its original trajectory. The ball is thus able to travel from one point to another in the same amount of time (as seen from the outside universe) via two trajectories.

During the weeks that Thorne was thinking about these problems, he was asked for a small favor by a friend, a recently retired engineer

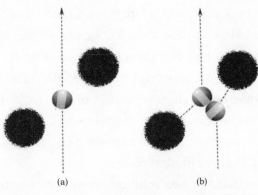

(a)   (b)

**Figure 5.4 Two Solutions for a Billiard Ball
and a Closed Timelike Curve**

named Robert Forward. Thorne had first met the older man in the 1960s, in order to interview him on his role in the development of the first laser interferometric gravitational wave detector.[27]

Forward was an unusual mix of experimentalist and theorist. He studied gravitational physics at the University of Maryland in the 1960s, and for his doctoral thesis he built and operated the first bar antenna designed to detect gravitational radiation. It is now in the Smithsonian. Forward worked at Hughes Aircraft Corporate Research Laboratories for thirty-one impressively productive years, authored hundreds of technical papers, and was awarded eighteen patents. He retired from Hughes in 1987, yet remained a consultant to NASA and the U.S. Air Force, working to develop "nonpropulsive drives"— among them lightsails, antimatter engines, and an electromagnetic tether (the famous "elevator to space"). He retired to devote more time to other activities, including writing science fiction. Although some readers said his characterization and style were juvenile, they acknowledged that his science was credible. In fact, Forward worked hard to stay current, and his query to Thorne was a case in point. It was precisely because he was deriving the plot of his work-in-progress from recent articles on closed timelike curves that he wanted Thorne to review his notes.[28]

An Infinity of Solutions

Thorne expected that the notes would offer a few hours of pleasant diversion. He had not expected them to supply a third solution to the billiard ball problem. It was this: The collision of the ball with itself need not occur when the first ball reaches the imaginary line between the mouths. It may occur earlier.[29] Thorne realized that if the ball emerging from the left mouth were given enough time to travel to the point of collision, the collision could occur earlier still. In fact, by using the closed timelike curve, the ball could buy all the time it needed.

Suppose that the right mouth is given three seconds worth of time dilation, and that, upon emerging from the left mouth, the ball is three seconds in the past as measured in the outside universe. Rather than traveling on a trajectory that would cause it to collide with its younger self, it takes one second to travel across the intervening normal space to the right mouth, enters the right mouth, and emerges again from the left mouth. Having made two trips through the wormhole, each carrying it three seconds past-ward, and having spent one second aging normally in the outside universe, the ball is now five seconds in the past as measured from the outside universe, and thus has five seconds to travel to the point of collision.

Suppose it needs still more time. It may enter the right mouth again and again, jumping backward in time (always by two-second intervals) an hour, a week, or a century, and so buying for itself more time to travel to a collision that could occur earlier and earlier, at a greater and greater distance from the mouths. Thorne realized that the number of possible trajectories for the billiard ball was not merely two or three. It was infinite.

Thorne and Klinkhammer suspected that classical Newtonian mechanics could not explain the infinite number of trajectories, and they reasoned that any understanding of the result could come only through another set of physical laws—those of quantum mechanics. To understand the problem facing Thorne and his student, and to appreciate the ingenuity of their solution, we must venture deeper into this realm.

The Strange World of Quantum Mechanics

The laws of classical Newtonian mechanics govern the universe on large, macroscopic scales—those of galaxies and stars and human beings. They are laws that we find reasonable, and that fit our intuitive sense of how things work. Quantum mechanics, on the other hand, governs the universe on very small scales—those of atoms and electrons and atomic nuclei. Its laws most decidedly do *not* fit our sense of how things work.

In his famous lectures at Caltech in the early 1960s, Richard Feynman introduced quantum mechanics by asking his students to surrender all presuppositions: "Things on a small scale behave like nothing that you have any direct experience about. They do not behave like waves, they do not behave like particles, they do not behave like clouds, or billiard balls, or weights or springs, or like anything that you have ever seen."[30] He then proceeded to describe the phenomenon that has become emblematic of what physicists call "quantum weirdness": the *double-slit experiment*.

Chapter 2 mentioned that the wave theory of light held sway through most of the nineteenth century. The theory was put on a firm footing in 1801 by English physician and physicist Thomas Young, who showed that when monochromatic light is passed through a pair of small apertures or slits and strikes a screen, it produces a pattern of light and dark bands. The pattern, Young noted, was exactly like that made by interfering waves on the surface of a liquid. Where two wave crests converge, the waves interfere with each other to produce an especially high wave crest; where two wave troughs converge, they produce an especially deep trough; and where a wave crest converges with a

Willem Jacob van Stockum during academic year 1934–35, two years before he demonstrated that Einstein's theory of general relativity allowed pastward time travel. (Courtesy of Christopher G. Oakley.)

Albert Einstein, whose theory of general relativity permits pastward time travel, with Kurt Gödel, who conceived of a model of a universe that would allow such travel everywhere, in Princeton, c. 1949. (Photograph by Oskar Morgenstern, courtesy of the Archives of the Institute for Advanced Study.)

Kip Thorne, who led the first teams to investigate the possibility of pastward time travel in the universe we know. (Courtesy of Kip S. Thorne.)

Igor Novikov (here enjoying some light reading), who proposed that pastward time travel need not violate causality. (Photograph by Tune Andersen.)

J. Richard Gott, who conceived a time machine of cosmic strings, wearing his "coat from the future" and holding a model of the universe giving birth to itself. (Courtesy of the Office of Communications, Princeton University.)

Stephen Hawking, who proposed that nature might prohibit pastward time travel. (Getty Images.)

Matt Visser, who has conceived numerous types of time machines, among them the "Roman Ring." (Courtesy of Matt Visser.)

David Deutsch, who argues that if pastward time travel is possible, it would be between whole universes. (Photograph by Daniel Kuan Li Oi.)

trough, they cancel each other out, leaving the surface relatively undisturbed.

Young hypothesized that his experiment had produced the same result, called an "interference pattern." The bright bands on the screen were produced by the convergence of two crests or two troughs, and the dark bands were produced by the convergence of a crest and a trough. Light, it seemed obvious, was made of waves.

In 1909, English mathematician and physicist Geoffrey Ingram Taylor demonstrated that even a feeble light source could produce such a pattern, and when Planck had shown light to be absorbed in "quanta" (the word means "discrete amounts") physicists began to wonder what might happen if quanta of light (the term "photon" came later), were sent through the apparatus one at a time. Classical Newtonian mechanics might predict that they would behave as particles, and that strikes on the screen would gradually accumulate to produce two patches of light, one behind each slit. But in the 1920s and 1930s, quantum theorists conducted a series of rigorous thought experiments and determined that this is *not* what would happen. Instead, an interference pattern would appear, and it would be the same interference pattern produced by Young's experiment. In the 1980s the double-slit experiment was actually performed, and results were exactly as predicted. It has been performed hundreds of times since, and it has produced an interference pattern every time.

It is worth pausing for a moment here to appreciate how strange such behavior is. How can a single photon contribute to building an interference pattern when, if it travels through the apparatus alone, there is nothing for it to interfere *with*? It is as though a given photon knows where others have struck the screen and adjusts its trajectory accordingly. Alternately, it is as though a force of some kind directs each photon to travel through one slit or the other.

Let us put ourselves in the position of physicists puzzling over the result. Although we have been sending photons through one at a time, we suspect that the photons are still somehow interfering with each other. So we change the experiment. We send one through, wait a day and send a second through, wait another day and send a third through. Eventually, to our astonishment, we see the same interference pattern.

Still puzzled, we return to the apparatus and make small changes. We know that, with one slit open, photons striking the screen accumulate to produce a patch behind that slit, and no pattern appears. We close the left-hand slit and send photons through the right-hand slit, and as expected, a patch appears behind it. Then we close the right-hand slit and send photons through the left-hand slit, and a patch appears behind it. We superimpose the patches, expecting that we might reproduce the interference pattern. But all we get are two superimposed patches, and no pattern.

The pattern, then, is produced only with two slits open simultaneously. When we realize that, the phenomenon becomes stranger still. It seems that the condition of the slit through which the photon is *not* traveling (that is, open or closed) affects the path of the photon through the other slit. On atomic scales, the distance between the slits is considerable, but the photon's behavior responds to information (the closed or open condition of the slit) that seems to be sent instantaneously. It is as though the photon "knows" whether the other slit is open or closed.[31]

Perhaps the behavior is explained by an odd quality of photons. Photons are, after all, a rather peculiar sort of subatomic particle: they are massless and travel always at the speed of light. So we substitute the photons with electrons—particles that possess mass and are known to behave much like classical objects. We perform the experiment again. It makes no difference.[32] In fact, the pattern has been produced with the entire menagerie of subatomic particles (now numbering over a hundred), as well as with entire atoms. The last, as physicist John Gribbin notes, are objects large enough to be photographed with an electron microscope and are approaching a scale with which we are familiar.[33] The quantum realm is not as removed as we might suspect (or as some might wish).

That quantum mechanics successfully predicted the results of the double-slit experiment comes as no surprise to most physicists, as the theory has proved its worth over and over. It has explained the structure of atoms, radioactivity, the effects of electrical and magnetic fields, the thermal and electrical properties of solids, and superconductivity. It has made possible technological developments like the laser, the transistor, and the electron microscope. In the sixty or so years since its beginnings, no experiment has contradicted its predictions. Indeed, it may be the most successful theory in the history of science. But for all this, the *meaning* of quantum mechanics remains controversial. It is worth remembering that the meaning of the laws of classical Newtonian mechanics, and the meanings of general and special relativity are self-evident. None of these require interpretation. But depending on how one divides them, there are at least five and as many as eight interpretations of quantum mechanics. The most accepted remains the Copenhagen interpretation, named for the city that is home to Niels Bohr's physics institute, where much of it was formulated.

The Copenhagen Interpretation

Central to the Copenhagen interpretation is the *Heisenberg indeterminacy principle*, which was put forth by German physicist Werner Heisenberg in 1927. Among other things, the principle asserts that, on quantum scales, we cannot measure the position and motion of an object at the same time. If we successfully measure an electron's momentum, we must forgo any knowledge of its position. If we successfully measure its position, we must forgo any knowledge of its momentum. It is important to appreciate that this inability is not caused by the clumsiness of our instruments. Rather, it seems that the electron *does not possess* a position and a momentum at the same time—a fact that has rather profound consequences for our understanding of the double-slit experiment.

A thrown baseball follows a well-defined path: it has position and momentum simultaneously at every point along that path. But because the particle traveling through the double-slit apparatus does not have a position and a momentum at the same time, it can have no definite trajectory. Although the experimenter may measure a point of departure and a point of arrival, she cannot infer a single, definite path between them. The question then arises, what exactly is going on between that departure and arrival? The Copenhagen interpretation would say that the electron did not exist, at least in the usual sense of the word. It was, in some sense, brought into being by the observation.

The *Copenhagen interpretation* then, asserts that physics cannot really answer the question of what an electron is. What it *can* do— and all we should expect it to do—is make a meaningful prediction about the behavior of a quantum system. Because the electron is encountered only when we attempt to measure it, a physicist interested in making correct predictions about its behavior must consider the electron only as a way of talking about the relations between observations. For his purposes it is not necessary that the electron actually exist independent of those observations.

Feynman's Sum-over-Histories Interpretation

Another formulation of quantum mechanics was developed by Feynman in the early 1940s, while he was a graduate student at Princeton working under John Wheeler. It claims that, to calculate the probability of the trajectory of a quantum entity—like that of the electron in the double-slit experiment—we must take into account *every possible trajectory*, including those that are slightly off the direct or classical path, as well as those that trace extremely lengthy and circuitous routes—via, say, Shanghai. These are probability waves. Like material waves, they are not in step with each other, and in fact interfere with each other. Also like material waves, probability waves possess amplitudes—that is, vertical distances from rest position to crest, or from rest position to trough.

As it happens, the longer and more circuitous paths tend to have amplitudes that cancel each other out, while paths nearer our idea of a classical trajectory reinforce each other. Feynman's formulation is termed "path integral" because to determine the probable path, all the paths are integrated, or "sum over histories" because the probable paths, here called histories, may be predicted by summing the amplitudes and squaring the total amplitude. There is an infinite number of possible histories, so the approach requires summing an infinite number. (This is not an insurmountable difficulty—indeed, using calculus, mathematicians perform such procedures regularly.) The sum-over-histories formulation explains the result of the double-slit experiment by claiming that, when an electron travels through the apparatus alone, it produces the characteristic interference pattern because it *is* interfering with itself— or more precisely—it is being directed by other histories.

This was an elegant solution, and it delighted Wheeler. He noted that the approach did not contradict classical mechanics so much as it included that set of laws within itself. In his autobiography, he tells of visiting Einstein on the off chance that he might use Feynman's work to convert the most famous skeptic of quantum mechanics. Einstein listened politely, and when Wheeler had finished, he replied, "I still can't believe that the good Lord plays dice." It was the rather vivid summary rejection of the theory that he had used for more than a decade.[34] This time, though, he added a note of bemused reflection: "Maybe I have earned the right to make my mistakes."[35]

In ensuing years a majority of physicists came to believe that Einstein's dismissal of quantum mechanics *was* a mistake, although a mistake he had earned the right to make several times over. Meanwhile, various interpretations of quantum mechanics were developed. By the early 1990s, when Thorne and his students were sending (imaginary) billiard balls through (imaginary) closed timelike curves, Feynman's sum over histories had many adherents, and it neatly explained their otherwise problematic results. Of the infinite number of trajectories possible, almost all would have very small probabilities, but there would be two trajectories with very large probabilities—and these were identical to the two self-consistent trajectories.

The Conclusions of the Consortium

For more than a year—roughly from the summer of 1989 to the summer of 1990—Thorne, Echeverria, and Klinkhammer had been conducting thought experiments involving closed timelike curves. Morris was working with John Friedman, who had also used the sum-over-histories interpretation to explain Polchinski's paradox. Yurtsever was still giving some of his attention to the subject, and Novikov was working on little else. Each was communicating with the others regularly, and each had a hold on a different part of what they were all increasingly beginning to appreciate as a single large question.

Thorne began to refer to the group as "the consortium." In June 1990, *Physical Review D* received a paper jointly authored by seven physicists.[36] It summarized the recent work on closed timelike curves and, building on the work of all seven authors, concluded that the sum over histories nicely complemented the principle of self-consistency. Novikov's self-consistency was thus provided a basis in quantum theory.

A Second Surprise

Earlier we said that a given billiards shot is difficult to replicate because of imperfections in the ball and small disturbances in its environment. The whole truth is that, if the instruments with which we measure speed, trajectory, and so forth were sufficiently sensitive, we would discover that, owing to quantum indeterminacy, the shot is *impossible* to replicate. The ball, after all, is made of atoms, whose exact position and momentum we cannot determine simultaneously, and which in fact do not exist simultaneously. Although these uncertainties affect the classical world and in fact underlie it, their effect on experiments with billiard balls is so small that, in both the actual, physical shot and its idealized mathematical description, they may be safely disregarded.

Quantum effects, then, are constrained by their scale. And so arose another unexpected finding from the work of Thorne and his students. It was a characteristic of time machines that was especially intriguing or—depending on one's point of view—disturbing. Two trajectories might be expected on the quantum scale, but Echeverria and Klinkhammer had found that, with time machines, two trajectories could be produced on classical scales. Time machines jumped the boundary from quantum to classical as if it did not exist. They magnified quantum uncertainty until it could be experienced on the macroscopic scales of billiard balls—and people.

In 1990, Novikov held a number of posts simultaneously—among them, head of the Department of Theoretical Astrophysics at the Lebedev Physical Institute in Moscow—but he was spending that academic year in a visiting position at the University of Copenhagen. It was during this stay that he suffered minor myocardial damage—a heart attack. Immediately he returned to Moscow, where doctors confirmed the diagnosis but admitted that both the tests and the surgery necessary to treat the condition were beyond the abilities of Soviet surgeons. They had ample cause for humility. Evgeny Lifshitz—one of the physicists who, in the 1960s, as we recall, proposed hypotheses of the nature of black hole singularities—had died as a consequence of such an operation.

Because Novikov could not afford the procedure in western Europe or the United States, his wife Eleonora wrote to Thorne asking for help. Within the month, Novikov was at Thorne's home in Pasadena, awaiting tests. The doctors discovered that his condition was more serious than first thought, and they made preparations to operate as soon as possible. There was, however, a difficulty with cost. Although the surgeon, cardiologist, and anesthesiologist at Huntington Memorial Hospital in Pasadena all refused payment, there remained ancillary expenses, and these were far beyond the means of Thorne. But he found another way. He contacted Novikov's colleagues worldwide and solicited funds. Soon enough, checks began to arrive, and a small sum began to accumulate. In due course the operation was carried out successfully. Novikov recalls coming out of anesthesia to the sound of Thorne's voice. Caltech's Feynman Professor of Theoretical Physics was saying, "*Tikhii, tikhii*"—"it's all right" in Russian.

Like many who come near death and escape it, Novikov won a renewed appreciation for life. But he gained something else as well—an experience of human compassion that his childhood under Stalin had worked to deny him. In 1998 he wrote, "The wonderful friendship of Kip and his wife (who make a movingly harmonious couple), the unselfishness, exceptional skill and attentiveness of American doctors, and finally, the brotherhood of the physics and astrophysics community, granted me a second life."[37]

Novikov was (and is) tireless, and during that rather eventful year he continued to write and publish, even as the world around him was reshaped. The two "time machine" papers that saw print that year gave his institutional affiliation as the USSR Academy of Sciences—an organization that, strictly speaking, would soon cease to exist.[38] During the year of Novikov's operation, the government that, for better and for worse, had determined the course of much of his life unraveled. In February 1990 the Soviet Communist Party agreed to surrender its power, and the constituent republics began to assert their sovereignty as superseding that of Moscow.

Novikov saw no reason to continue to live in the place again called Russia, and because he was a man of extraordinary talents and accomplishments, he was not long between jobs. Officially he retained his position at the Lebedev Physical Institute, but by early 1991, when the leaders of the Russian, Ukrainian, and Belarusian republics declared that the Soviet Union had been replaced with the Commonwealth of Independent States, he had settled his family in Denmark, where he had accepted the offer of a permanent position as a professor of astrophysics at the University of Copenhagen.

In 1990, Novikov and Russian physicist Valeri Frolov published a coauthored piece. Its title, and indeed the titles of Novikov's subsequent publications on closed timelike curves, dropped all pretense of coyness. In fact they became almost startlingly frank. The article written with Frolov was called "Physical Effects in Wormholes and Time Machines."

The paper's content was every bit as audacious. Frolov and Novikov suggested that moving a wormhole mouth at relativistic speeds might not be the only way to slow its aging. "Suppose," they wrote, "that one of its mouths moved close to the surface of a neutron star and it is held there at rest by some external force while the other mouth remains far from the star."[39] The mouth near the surface of the star would age at a slower rate, and that rate would be determined by several factors, including the distance from the surface and the star's mass. If the neutron star were average-sized and the mouth were on its surface, the mouth would age three hours for every ten hours in the outside universe. After the passage of, say, ten years in the outside universe, our suitably advanced civilization could retrieve the "preserved" mouth and return it to the vicinity of the other mouth. In so doing, it would have created a time machine that would allow travel seven years into the past.

In the same paper, the authors described a means to see inside a black hole's event horizon, a place that almost by definition was thought forever beyond our ken. They suggested that, by dropping one mouth *below* the event horizon, the suitably advanced civilization would be able to see its interior through the other mouth, the wormhole having become a sort of remote-controlled camera. Stranger still, because a wormhole allows the transport of material bodies as well as light, it would be possible to use it as a lifeline, lowering one mouth below the horizon to rescue an observer unlucky enough to have fallen through, or (although Frolov and Novikov did not suggest this) to remove an especially brave explorer.

Intriguing as such speculation was, it was only a digression. The larger purpose of Frolov and Novikov's 1990 paper was to make Thorne and Morris's wormhole model of a time machine more realistic. Indeed, that model had been extremely idealized—or more precisely, its environment was extremely idealized; the wormhole was floating in space beyond the reach of any external forces.

Frolov and Novikov took the same wormhole and placed it in a more realistic setting—one subject to gravitational and magnetic fields. They did not need to specify a location. Virtually every point in the universe is affected by gravity, and even a weak gravitational field will cause an acceleration that will steadily increase. Because the mouths occupy different positions in space, they will be affected differently by gravitating masses. Inevitably, one mouth will be accelerated more than the other, and the differences between their relative speeds, which might begin as a mere microsecond, would grow larger. Their reference frames will grow farther apart. What this means, in the words of the authors, is that "almost any wormhole placed in an external gravitational field or interacting with external matter becomes a time machine."[40]

CHAPTER SIX

COSMIC STRINGS
AND CHRONOLOGY PROTECTION

IN WHICH ONE PHYSICIST PROPOSES A NEW AND INGENIOUS MANNER
BY WHICH ONE MIGHT TRAVEL PASTWARD, AND ANOTHER
CONTENDS THAT NATURE FORBIDS PASTWARD TRAVEL
BY THIS OR ANY MEANS WHATSOEVER

Could time be reversed, and the future change places with the past, the past would cry out against us, and our future, full as loudly, as we against the ages foregone. All the ages are his children, calling each other names.
— Herman Melville, *Mardi, and a Voyage Thither*

In the spring of 1988, when Thorne was in Wisconsin working on the first time machine paper, he took advantage of the relative proximity of two colleagues at the University of Chicago: Robert Geroch and Robert Wald. Both were members of a general relativity research group formed by Chandrasekhar,[1] and Thorne expected that both would have an interest in closed timelike curves.

Geroch was among the physicists who, in the 1960s, had developed "global methods," a way to perform general-relativity calculations without using Einstein's field equations directly. In 1967, while he was a graduate student (like Thorne, he had worked under Wheeler), Geroch had published an article proving that a certain type of wormhole could be constructed only by movement of its parts backward and forward in time.[2] The paper caused a small stir at the time because it raised the specter of causality violation, but like other models previous to 1988, it was thought unphysical and no cause for real concern. Wald had also been a student of Wheeler. He had long been interested in quantum effects in the vicinity of black holes.

One day in March, Thorne presented his thinking on closed time-like curves to an audience that included both men. By that time, Thorne's wormhole/time machine model was fairly sophisticated, but the universe in which it lived was still highly idealized, a mostly empty stage. One feature of our actual, nonidealized universe is the electromagnetic radiation produced by stars, quasars, and the background noise of the big bang. Geroch and Wald reminded Thorne that it would be present naturally around and inside any actual wormhole. More than that, they suggested, it might well thwart any attempt to turn the wormhole into a time machine.

New Doubts

Suppose we wanted to create the billiard ball time machine from the previous chapter—that is, a wormhole whose right mouth has been subjected to one second of time dilation relative to the left mouth (and, of course, the rest of the universe). We would accelerate the right mouth away from the left mouth at nearly the speed of light for a little more than half a second, then accelerate it in the opposite direction again for half a second, returning it to the vicinity of the left mouth.

Before we begin to turn the wormhole into a time machine, however—that is, before we accelerate the right mouth—we observe that ambient electromagnetic radiation is plentiful in the vicinity of both mouths of the wormhole and is coursing through the throat in both directions. Some radiation—a small but finite amount—travels from the vicinity of the stationary mouth via normal space to the vicinity of the moving mouth. Because this radiation is traveling at the speed of light, it is somewhat faster than the moving mouth and can overtake it. Some of this radiation—again, a small but finite amount—enters the moving mouth, makes the trip through the throat, and emerges from the stationary mouth.

As long as the moving mouth is outward bound, nothing unusual happens—which is to say, the radiation emerging from the stationary mouth emerges *after* it left. But sometime after the outward-bound mouth reverses direction, the radiation begins to emerge from the stationary mouth *at the same moment* it left. It then adds itself to the part of the radiation in the vicinity of the stationary mouth that travels via normal space to the moving mouth, some part of it enters the moving mouth, and so on.

That the radiation emerging from the stationary mouth does so in ever-diminishing amounts does little to offset the larger effect: an avalanche of radiation. Moreover, because the moving mouth on its inward-bound leg would be traveling at a significant speed, the radiation is greatly blueshifted (that is, squeezed into shorter wavelengths) and so carries far more energy than it otherwise would. According to Morris and Thorne's teaching paper, it is at some moment after the moving mouth reversed direction that the wormhole is turned into a time machine. But, Geroch and Wald argued, it is also at this moment that ambient electromagnetic radiation would destroy the wormhole.

Thorne was caught off guard, but only for a few hours. As he drove back to Wisconsin, he thought about the behavior of radiation. He realized that, although Geroch and Wald were correct to suggest that it would travel unhindered through the throat, they had overlooked a consequence of the shape of space in the vicinity of a wormhole. An embedding diagram of that space might look like two trombone bells facing in opposite directions and joined at the narrow part of their throats. Electromagnetic radiation would follow geodesics along the surface of one bell and into the throat, where it would begin to converge. At the point the throats were joined it would begin to *diverge*, and it would continue to diverge as it followed geodesics along the surface of the other bell. Thorne could calculate the energies as he was driving; he convinced himself that this divergence, or "defocusing effect," would prevent the emerging radiation from building in intensity to a point at which it was powerful enough to destroy the wormhole. He was willing to concede that there could well be forces in nature that prohibited closed timelike curves, but radiation was not among them.

Thorne assumed that he and Morris and Yurtsever had addressed the more "material" obstructions to closed timelike curves. He had put such concerns aside to wrestle with Polchinski's paradox. Still, he had a gnawing suspicion that he was missing something, and as was his habit, he sought the views of others. One of these was William Hiscock, who had done work in marrying the very small with the very large—applying the physics of elementary particles to relativistic astrophysics. Thorne told Hiscock of his doubts, and Hiscock suggested that he consider the effect of electromagnetic vacuum fluctuations.

Vacuum Fluctuations

If one could take a region of space and somehow remove from it all particles and fields, until one had achieved the purest possible vacuum, that vacuum would still be seething with fluctuations caused by the momentary give-and-take of energy between bordering areas. "The dynamic vacuum," writes Hans Christian von Baeyer, "is like a quiet lake on a summer night, its surface rippled in gentle fluctuations, while all around, electron-positron pairs twinkle on and off like fireflies."[3] The *wave–particle duality* allows us to view these fluctuations as waves or particles. If we choose to regard them as waves, their form moves in ways that are random and unpredictable, with positive energy in one place, negative in another. If we choose to regard them as particles, they appear in pairs, live for a brief moment on energy borrowed from a nearby area, and then annihilate each other. We may call them "virtual" because they are extremely short-lived and represent an intermediate stage in a larger process.[4]

Thorne's first thought was that, much as the defocusing effect of the wormhole mouths would diffuse radiation, so it would diffuse vacuum fluctuations. This view, he found, was shared by others in the consortium, among them Valeri Frolov. Still, he had questions. So with guidance from a 1982 article by Hiscock and mathematical physicist Deborah Konkowski,[5] Thorne and Caltech postdoc Sung-Won Kim set out to predict exactly how vacuum fluctuations coursing through a wormhole/time machine would behave. After a year of work on the problem they had an answer, and it surprised them.

Imagine the wormhole at the moment it becomes a time machine. Vacuum fluctuations travel at the speed of light through normal space to the moving mouth, course through the throat, emerge from the stationary mouth and begin to diffuse, as does radiation. But then they begin to behave quite unlike radiation. They regather themselves, refocus and travel through normal space to the moving mouth, enter it, course through the throat, and emerge from the stationary mouth *at the same time* they left. They repeat this circuit again and again, becoming infinitely intense, more than powerful enough to destroy the wormhole.

It appeared that the process that Thorne had proven would *not* operate with radiation *would* operate with vacuum fluctuations. At first, this seemed a serious, perhaps insurmountable obstacle. Vacuum fluctuations are part of the fabric of space; they cannot be shielded out. But Kim and Thorne found that the wormhole might yet survive.

When they graphed the evolution of the fluctuations, they found that they spiked immediately, rising to an intensity sufficient to destroy the wormhole, continuing to rise to infinite intensity, and then falling to a level below that sufficient to destroy the wormhole. It was like a siren rising and falling between two ticks of a metronome, except that the interval between the ticks would be unimaginably brief—about 10^{-43} of a second. Recall that this is a Planck-Wheeler time, the smallest interval of time that exists, the interval smaller than which concepts of "before" and "after" are meaningless.

Within a single Planck-Wheeler time the vacuum fluctuations could not build to infinite intensity; in fact, they could not build at all, because building is a process that requires befores and afters. Kim and Thorne reasoned that the spike could not be real. Rather, the actual evolution would be suspended at the beginning of the Planck-Wheeler time and allowed to continue only after its end. Thus the wormhole/time machine was saved from destruction by the speed of evolution of the fluctuations and the discontinuous, "granular" nature of time on the smallest scales.

As Thorne and Kim worked, a physicist on the East Coast named J. Richard Gott was thinking about another sort of time machine.[6] Gott has been a full professor of astrophysics at Princeton since 1987. He has a laugh that comes unexpectedly, and he speaks with long Kentucky vowels that seem at odds with half a lifetime lived in central New Jersey. More than once, Princeton students have voted Gott the school's outstanding professor. It is not hard to see why. His enthusiasm for physics is contagious.

Having read and admired the work of Morris, Thorne, and Yurtsever, Gott was inspired to imagine another means to accomplish the same ends. He conceived of a time machine that would need no wormhole and no exotic matter, but it would be strange in its own right. Its most crucial components were bizarre fossils from the very early universe called *cosmic strings*. To understand cosmic strings and how they might be employed to travel pastward, we will need to understand the conditions, at the very beginning of the universe, that allowed their formation.

The Beginning of the Universe

The big bang theory of creation, which posits that the universe began as a singularity that rapidly expanded in a "cosmic fireball," was first proposed in the 1930s by Georges-Henri Lemaître and developed in the 1940s by George Gamow and his students Ralph Alpher and Robert Herman. We should not think of the big bang as an explosion. An explosion needs a space outside it to explode *into*, and the big bang, occurring everywhere in the universe, had nothing outside it. We can envision it better by appealing to Riemannian geometry, and by imagining our three-dimensional universe as the two-dimensional curved surface of a sphere.

When the universe was young, this sphere was small, its surface a hot dense plasma of quantum particles; as the universe aged, this surface stretched and cooled, and the particles on the surface formed atoms, fundamental molecules, and eventually nebulae, stars, planets—and theoretical physicists. Gamow, Alpher, and Herman predicted that, because there is no "outside" the universe—that is, nowhere "off" the surface of the sphere—the heat of the big bang, having nowhere to go, would simply be dissipated on the greatly expanded surface. It would appear as very low-temperature radiation coursing everywhere through space.

In 1964, radio astronomers Arno Penzias and Robert Wilson, working with a large antenna at Bell Labs in Holmdel, New Jersey, detected a peculiar hissing sound. They first thought they were hearing noise from their electronics, but soon realized they had detected the radiation that Gamow, Alpher, and Herman had predicted—what would come to be called the cosmic microwave background. Then, in the early 1990s, NASA's *Cosmic Background Explorer* satellite emphatically confirmed the physicists' more specific predictions—that the radiation would have the same intensity everywhere in the sky, and that its temperature would be about three degrees Kelvin.

For all its eventual success, the big bang theory was beleaguered with three "problems": (1) the homogeneity problem (why matter on a large scale is distributed evenly across space), (2) the magnetic monopole problem (why a sort of topological defect predicted by the classical big bang theory did not slow and reverse the universe's expansion shortly after its beginning), and (3) the flatness problem (the topological flatness of the universe we measure would seem possible only if its initial rate of expansion were improbably precise). Almost as problematic was the theory's limited scope. It did not assign a cause to the big bang, and it had nothing to say about what might have happened *before* the big bang, if indeed there could be said to have been a "before."

One answer to these questions was suggested by physicist Edward Tryon in a 1973 paper that contained this rather memorable sentence: "I offer the modest proposal that our Universe is simply one of those things which happen from time to time."[7] He was proposing that the big bang began as a quantum fluctuation, and that this quantum fluctuation, despite our quite understandable fondness for it, was like all the others—that is, a purely random event. Tryon's idea was met with skepticism, not because physicists were disinclined to accept a genesis produced by chance, but because none could imagine how something so small could become so large, so quickly. But about 1980, a young particle physicist named Alan Guth, building upon the idea, began to advocate a modification of the big bang theory. Much could be explained, Guth argued, if at a moment very early in its existence, the universe grew very large, very quickly—expanding at an exponential rate by a process called *inflation*.

Guth's proposal has since gained in force and credibility, largely because it answered the three problems of the big bang theory. The first two answers, described in numerous works, will not concern us here. But the answer to the flatness problem bears directly on our subject.

The nineteenth-century mathematicians discussed in Chapter 2 determined the distinguishing characteristics of Euclidean geometry. They did not, however, determine whether the universe we live in is Euclidean, or, to use the more common term, flat. But in the twentieth century, astronomers found a new way to address the question. They realized that there must be a critical density of matter for the universe. If the density were greater than this, the collective gravity of all matter would slow and eventually reverse the expansion of the universe, causing it to collapse. This would be a *closed universe*. If the density were less than critical, the universe would continue to expand forever. This would be an *open universe*. Finally, if the density were precisely at the critical point, the universe would continue to expand ever more slowly but never quite stop its expansion. This would be a *flat universe*.

By the late 1970s, most observations supported a number very near the critical density for a flat universe. This result raised new questions. Of the three universes under consideration, a flat universe seemed the least likely. A flat universe could occur only if the forces of expansion and collapse were balanced with fantastic precision, yet physicists could imagine no reason those forces should be balanced at all. This was a very curious state of affairs, and although the big bang theory could not explain it, Guth's inflation offered an answer.

The classical big bang theory had predicted that the universe was the size of the visible universe or a bit larger. Inflation predicted that the present universe was perhaps one hundred times larger than the visible universe. Palm a basketball, and you can feel its curvature. Imagine that, as you continue to palm it, it is slowly inflated. When it doubles in size once, it feels flatter; when it doubles in size twice, its feels flat. Regardless of the shape of the young universe—sphere, saddle, or something in between—double its size a hundred times and even a region as large as twenty billion light-years across will seem flat. Inflation meant that suddenly there was no reason to require that the forces of expansion and collapse be balanced. In short, inflation solved the flatness problem by dismissing it.

Inflation and the False Vacuum

Let us put Guth's inflation in context and review what most physicists believe occurred during the very early history of the universe. The first 10^{-43} second of the universe was an unimaginably thin slice of time called the *Planck epoch*. Physicists can only speculate as to its nature. They believe that the universe's density was equal to the *Planck density*, but this means little more than that quantum effects were important. They suspect also that the four fundamental forces of nature—the gravitational force, the strong nuclear force, the weak nuclear force, and the electromagnetic force—were united in a single "superforce." But they are uncertain even of this. A better understanding of the Planck epoch must wait for a theory of quantum gravity.

The end of the Planck epoch was marked by the decoupling of the gravitational force from the other three forces, which remained indistinguishable. This began the "epoch of grand unification." Although, like the Planck epoch, the epoch of grand unification lies beyond reach of direct experiment (the universe's temperature was 10^{28} K, far higher than temperatures achievable by particle accelerators), it is within the reach of theory—specifically, the theories developed in the 1970s that unified three of the four forces. These are termed *grand unified theories*, or GUTs.

Ideas of inflation, as presented by Guth and others, suggest that during the epoch of grand unification, space itself expanded, doubling in size every 10^{-35} second and repeating this doubling at least a hundred times before, an unimaginably brief interval later, it ceased. Then, when the universe was a mere 10^{-33} second old, its rate of expansion slowed to the leisurely pace at which it continues today. By comparison, this rate is barely a crawl; current estimates are that the universe will need ten billion years to achieve another doubling in size. No matter; the important work was done. As we may recall from a grade school introduction to exponents, if we take a penny and double it every day, in a month we produce a rather large sum. In such a fashion, inflation had produced a very large universe very quickly.

The engine driving inflation was something called a *false vacuum*. Exactly how might we understand a vacuum to be "false"? Recall that the vacuum is not empty; quite the contrary, it is seething with the activity of electromagnetic particles being created and annihilated. In a "true" vacuum the energies cancel each other out; in a false vacuum they do *not* cancel each other out. It is for this reason that a false vacuum has more energy than does a true vacuum. But that energy, and the false vacuum itself, will not survive long. The false vacuum is unstable. Or more precisely, it is *metastable*: after a brief interval it decays into a true vacuum.

There is no reason to expect this decay to progress everywhere at the same rate. Much like the boiling water in a pot suddenly removed from a burner, certain regions cool faster than others, and like bubbles in the water, bubbles of false vacuum disappear when their pressure falls below that of the medium surrounding them. Of course, the surrounding mediums at issue here are different; the water exerts a measurable pressure, and true vacuum exerts no pressure. Strangely, though, the true vacuum still overwhelms the bubbles of the false vacuum. This can only mean that the pressure exerted by those bubbles is less than zero. And so it is. In fact, the false vacuum may be imagined as a sort of suction, with the important qualification that a suction is created when a pressure in a fluid drops below that of a surrounding fluid. The false vacuum that concerns us here fills the entire universe, and there can be nothing surrounding it because (again) there is no "outside" the universe.

If the false vacuum acts as a suction, we might expect it to slow the universe's expansion. But, quite the contrary, it accelerated it.

How? In Newtonian physics gravitational forces are produced only by masses. But as Einstein's $E = mc^2$ shows, mass can be described in terms of energy. One sort of energy is pressure, and it follows that pressure can actually generate a gravitational field. The fields produced by pressures familiar to us—atmospheric pressure, for instance—are so weak as to be negligible. But during the first moments in the life of the universe, all pressures were extreme and the fields they generated were significant. They were also varied. As a positive pressure produces a gravitational field that attracts, so a negative pressure produces a gravitational field that repels. (We should note that, although this is negative pressure, it represents positive energy and should not be confused with the negative energy that Morris and Thorne suggested might hold open a wormhole.)

Guth argued that, as the false vacuum drove inflation, the universe's average temperature dropped dramatically. This rapid cooling is what allowed the strong nuclear force to decouple from the weak nuclear force and the electromagnetic force—an event that marked the end of the epoch of grand unification. The decoupling is called a "phase transition"—a dramatic change, somewhat like the transition of water to ice, that is characterized by a loss of symmetry.

Examining a drop of water under a microscope shows that every part of it looks like every other part. It has symmetry. Freezing the drop into a thin sliver of ice turns it into a crystal, with molecules arranged in lattices oriented in different directions. Every part no longer looks like every other part. Moreover, because the water began to freeze in many places independently, the plates of ice joined in a random fashion, leaving boundaries between them. These are what physicists call "topological defects."

The Formation of Cosmic Strings

The phase transition that ended the epoch of grand unification may also have left topological defects, inside which material from that epoch was trapped. One sort, described in 1976 by physicist Thomas Kibble, would be very long, very dense filaments called *cosmic strings*. They would be extreme phenomena indeed. Because they have no ends, cosmic strings would appear in two versions: those that stretch across the visible universe, and those that form loops. Neither version would be directly visible, as a cosmic string in cross section would have a diameter smaller than that of an atomic nucleus. But both would be remarkably massive: a length of 1.7 kilometers would weigh more than Earth. (Cosmic strings should not be confused with the strings of string theory, the hypothesized fundamental entities with lengths of 10^{-33} centimeter and effectively zero width.)

We might expect so much mass concentrated in so small a volume to produce a significant gravitational field, and indeed it would. But as cosmic strings would preserve inside themselves the mass density of the epoch of grand unification, so they would also preserve the false vacuum of inflation. As it happens, the attractive gravitational field of the strings' mass would be exactly offset by the repulsive gravitational field of the false vacuum. Consequently, no gravitational forces would be felt outside a cosmic string. Strange as this feature seemed, physicists were to discover others that were stranger still.

In 1985, theoretical physicist William Hiscock and J. Richard Gott independently found exact solutions to Einstein's field equations for cosmic strings. Both discovered that a cosmic string running at right angles through the center of a circle of flat space would change it dramatically. It would be as though a narrow slice of the circle had been removed and the rest pulled downward, and the two edges where the slice was missing had been drawn together and sewn shut. The circle was turned into a cone. For a "realistic" cosmic string— that is, one with the features predicted by Kibble—the circumference of the cone would be smaller than the circumference of the circle by only one ten-thousandth of a degree. This is a very small difference to be sure, but large enough to create some curious effects.

Suppose we begin again with the circle of flat spacetime. Now, though, we imagine that the region of the circle is crossed by two parallel geodesics of the same length—one on either side of the circle's center point—and that the right one passes at a greater distance from the center than does the left one (Figure 6.1a). When we remove the slice, pull the circle downward, and sew together the edges where the slice is missing, we cause the geodesics to intersect (Figure 6.1b). We also decrease the length of the right geodesic. This means that light following the right geodesic would not have as far to travel as light following the left geodesic, and would arrive at the intersection sooner. Perhaps more intriguingly, a traveler following the right geodesic, if he were fast enough, could actually win a race against light.

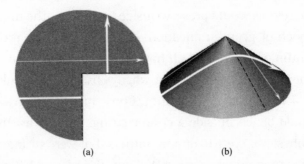

(a)   (b)

Figure 6.1 The Effect of a Cosmic String upon Two Geodesics

To illustrate, let us suppose that a cosmic string threads interstellar space halfway between Earth and Arcturus, and that Gassendi and his assistant agree to race around it. Gassendi travels along the "normal" geodesic, and although his ship is very swift—indeed it travels within a fraction of the speed of light—Gassendi cannot arrive at Arcturus sooner than thirty-six years after he left. (This duration, of course, is as measured in the outside universe; to Gassendi himself the trip could be fairly brief.) Gassendi's assistant begins at the same place and time, but travels along a shortened geodesic. Although his speed as measured by a stationary observer is no greater than Gassendi's, he arrives at Arcturus well ahead of Gassendi. In fact, if the geodesic he travels is shortened enough and his speed is great enough, Gassendi's assistant could arrive at Arcturus before light that had begun at the same time and place but followed a "normal" geodesic. He could then look back along that geodesic and, with a sufficiently powerful telescope, actually see himself departing Earth.

Now suppose Gassendi has a second assistant, and before the race began she set out toward Arcturus along a normal geodesic. At some point in her journey she would have been able to look back toward Earth and (again with a sufficiently powerful telescope) see Gassendi's first assistant departing and, at the same moment, look toward Arcturus and see Gassendi's first assistant arriving.

Pastward Time Travel Using Cosmic Strings

In 1991, Gott realized that two cosmic strings moving past each other at high speeds would, according to special relativity, create in spacetime a kind of fault line, and that this fault line might be exploited for pastward travel.

Suppose that halfway between Earth and Arcturus we introduce a *second* cosmic string, and imagine that it is moving past the first, and the first is moving past it, at very high speeds. (If ordinary masses of equivalent density were to pass so near to each other, they would collapse into each other to become a black hole, but the absence of gravitational forces in the vicinity of the strings allows them to pass unimpeded.) If Gassendi's first assistant travels along a shortened geodesic, then, as before, the second assistant traveling a normal geodesic would at some point be able to see the first assistant both departing Earth and arriving at Arcturus at the same time.

However, Gassendi's first assistant does not stop at Arcturus. Instead, via another shortened geodesic he loops back around the strings when they are passing each other. In so doing, he takes advantage of the relative motion of the strings' reference frames. The first moves against the direction of travel of Gassendi's first assistant as he is outbound from Earth; the second moves against his direction of travel as he is inbound to Earth, and each moves against the direction of the other. If Gassendi's first assistant times his trip very carefully, he can return to Earth at the moment he left. (This action would produce a number of strange circumstances like those described in the previous chapter. For one, that he meets his earlier self upon return would mean that he met his later self upon departure, and so learned that his trip would succeed before he began it.)

By adjusting his speed and his moment of departure, Gassendi's first assistant might return to Earth at moments slightly later than his moment of departure, and he might choose any moment of return right up until that which he could accomplish with no time machine. As with the wormhole time machine, the cosmic-string time machine does not allow travel further pastward than the moment it was made—which in this case would mean the moment the strings' motion allows the path that Gassendi's assistant takes.

Gott admitted that manufactured cosmic strings would pose a challenge even for a sufficiently advanced civilization. The only laboratory that could hold infinitely long strings was the universe itself, and strings of finite lengths—that is, strings existing in a finite space—would collapse into themselves. Initially, Gott believed he had found a way to manage the problem. Because the loops need not be perfectly circular, he thought they might be given parallel lengths long enough and straight enough to serve the purpose. But he found that a straight section of sufficient length would require the whole loop to be so massive that, according to Thorne's conceivable conjecture, it would implode into a black hole. This did not mean that a natural cosmic-string time machine was impossible. Because cosmic strings were predicted by several grand unified theories, Gott gave their existence better than even odds.[8] It was conceivable that, somewhere in the history of the universe, two strings had passed each other at great speeds, thereby creating (if momentarily) a natural time machine.

Gott's ideas, which appeared in print for the first time in *Physical Review Letters* in early March 1991, were noteworthy for two reasons. First, they described a closed timelike curve made with no exotic matter but constructed entirely of materials that, although probably not nearby, were widely believed to exist. Perhaps more significantly, they offered a second example of a time machine. Two sightings of an animal thought to be mythical or extinct are more likely to persuade a zoologist of the animal's existence than is one sighting. If the second sighting reports some of the features reported in the first sighting and some features that are new, a zoologist might be prepared to entertain the possibility of an entire class of the animal. Gott's paper had suggested the possibility of an entire class of time machines.

Soon, the very week the paper appeared Thorne was visiting Princeton to speak about closed timelike curves. Gott took the opportunity to present his ideas of string-driven machines in person, using paper cones. Thorne was intrigued—so much so that, upon returning to Caltech, he discussed the idea with his research group. One of its members, Curt Cutler, began to embellish Gott's concept, and by late summer he had determined with considerable precision the region around the strings in which time travel would be possible.[9]

Soon the lay press began to show interest. In early May, *Time* magazine ran a full page on Gott's idea, complete with a diagram of a cartoon rocket looping through spacetime, and a photograph of a beaming Professor Gott, the fingers of one hand pinching the ends of a string, those of the other holding a toy spacecraft that had been a gift from his seven-year-old daughter. The article noted, "There must be two strings . . . for the trick to work. But work it apparently does."[10]

Others were not so convinced. On the same day that the *Time* piece appeared, Gott was in Boston, where he spoke to a group of physicists that included Alan Guth. In subsequent weeks, Guth and two colleagues grew convinced that Gott was in error. Although each cosmic string was moving at less than the speed of light, it was a mistake to think about them as separate from each other. Rather, they were properly regarded as parts of a system, and the "total momentum" of that system would, for all intents and purposes, be moving faster than light. Because such motion was an obvious impossibility, it effectively undermined the whole concept of cosmic-string time machines. (At about the same time, another group reached the same conclusion by a slightly different approach.) It was a measure of the quickening pace of the discussion that before the piece by Guth and his coauthors went to press, they added a qualification that their refutation would be valid only in an open universe. In a closed universe, they admitted, a cosmic-string time machine was possible. Clearly, possibilities were still in play.

Meanwhile, there were developments on another front. About eight months earlier, in September 1990, Kim and Thorne had worked through the problem of quantum fluctuations and concluded that they would not destroy the time machine. They sent a draft manuscript to *Physical Review D* and circulated copies among colleagues. One of these was Stephen Hawking, in the Department of Applied Mathematics at the University of Cambridge.

Hawking

In 1963, while a graduate student at Cambridge, Hawking was diagnosed with amyotrophic lateral sclerosis (ALS). On several occasions he has acknowledged that certain of his insights may be owed indirectly to that disability, as it has freed him from teaching and administrative responsibilities. He is afflicted with a particularly slow-working strain, and thus his physical decline has been gradual, allowing him time to adjust and to develop his own unique—and uniquely powerful—approach to problems. He does not work by scribbling equations on a blackboard in the manner of most theoretical physicists, but rather by creating and manipulating mental images. In 1985 he underwent a tracheotomy so that his lungs might remain clear of fluid; for years thereafter he communicated through a voice synthesizer that he operated with his hand. (More recently, he has replaced the hand controls with a device that tracks his eye movements.) Because the mind is quick and the machine is slow, the sentences that issue from the synthesizer are brief and, in their own way, perfect. It is these qualities that strike a listener, especially one in awe of Hawking's insights into cosmology, as oracular. But, as anyone who has heard his lectures will attest, he is possessed of a particularly dry sense of humor. The audience of those lectures, and even those who have known him for decades, are not always sure who is speaking at any given moment—Hawking the oracle or Hawking the jester. Sometimes— quite often, in fact—it is both.

Hawking and Thorne are near contemporaries (Hawking is two years older), and they came of age professionally during the 1960s and early 1970s, during the heyday of black hole research. Their friendship has an element of playful competition, manifest most famously in a wager made in 1974 concerning a radio source called Cygnus X-1. Thorne bet that it was a black hole; Hawking, that it was not. Sixteen years later the evidence was overwhelmingly in Thorne's favor, and Hawking conceded in June of 1990— as it happened, a few months before he received Kim and Thorne's draft article.

Hawking had not studied closed timelike curves specifically (in 1990, few had), but he had given considerable thought to extreme cases of spacetime curvature. In 1964, when Penrose had used the methods of topology to show that a star could collapse to a singularity even if it was not exactly spherical, Hawking began to use a similar method to imagine an analogous (but time-reversed) situation: the expansion of the universe from a singularity. The theorems associated with the method required him to think about how an event in spacetime influences points in its future—a process called "Cauchy development" after the French mathematician Augustin-Louis Cauchy. Hawking realized that the influence of any event is limited to a certain region, and he called the outer edge of this region a *Cauchy horizon.*

As he read Kim and Thorne's piece and began thinking about closed timelike curves, Hawking realized that they, too, must have a Cauchy horizon. In their case it has come to be termed a *chronology horizon*, defined as the boundary separating a region in which time travel is possible from a region in which it is not possible. Any closed timelike curve—whether created by a Morris-Thorne wormhole or Gott's cosmic strings—would create a chronology horizon.

As Hawking considered the piece further, he discovered what he believed was an error. Recall that the moving mouth of the wormhole time machine is traveling at nearly the speed of light, and at such velocities relativistic effects become significant and reference frames become important. The reference frame that Kim and Thorne had chosen was that of a hypothetical observer at rest in the wormhole throat; using this frame, they had concluded that the vacuum fluctuations would not have time to build to infinite intensity and destroy the wormhole. Hawking, however, believed that the proper frame of reference was that of a hypothetical observer coursing through the wormhole along with the vacuum fluctuations. The shift to this frame changed everything. It meant that the fluctuations would have sufficient time to destroy the wormhole.

The Chronology Protection Conjecture

In a seminar at Cambridge, Hawking addressed the problems he perceived in Kim and Thorne's model, and he suggested something else: that the destruction of the wormhole by quantum fluctuations was merely one of several means by which the universe protected itself against causality violation. In September 1991 it was an idea only partially formed, and there was little hard evidence to support it. Hawking had, however, already given it a name. He called it the *chronology protection conjecture.*

A *hypothesis* is a tentative explanation for a phenomenon that is supported by observational evidence—even if, in the case of theoretical physics, such evidence may be derived from a thought experiment. A set of hypotheses go to make a *theory.* (As Hawking himself has noted, a theory can never be proved, since we will never know everything there is to know about the universe. It can, however, be tested and so gain credibility over time. Of course, it can also be disproved.) Occasionally, a physicist will put forth an educated guess for which there is no evidence, only a gut feeling. This is termed a *conjecture.* Because Hawking had evidence for several mechanisms that prohibited closed timelike curves, his use of the term "conjecture" may have been unduly modest. The rest of the name, however, was anything but. The words "chronology protection" suggested an overarching natural law. The leap from doubt regarding the viability of a wormhole to a natural law protecting causality was considerable, yet Hawking had made such leaps before. His recent concession on Cygnus X-1 notwithstanding, Hawking's sense of how the universe behaved was unequaled.

Hawking described his conjecture again during an address in Kyoto in late June 1991.[11] In subsequent weeks he developed the idea into an article, and in September 1991 he submitted it to *Physical Review D*. In eight closely reasoned pages, Hawking proposed that any time machine imaginable would of necessity employ a closed null curve. Quantum fluctuations would travel along that curve, making circuit after circuit at the same time, and grow to a critical intensity. At the moment the warped space became a time machine—or rather, because quantum indeterminacy prevents the chronology horizon from being a sharp-edged boundary, a very brief interval *before* that moment, it would be destroyed.

On the last page of the paper, Hawking outlined two scenarios by which one might create a closed timelike curve, and he claimed that both, for different reasons, would fail. In the voice of the oracle, the paper concluded, "Either way, there seem to be theoretical reasons to believe the chronology protection conjecture: *The laws of physics prevent the appearance of closed timelike curves.*" Following this was a sort of coda that one suspects the editors of *Physical Review D* would have excised had the submission come from another author. It was a single sentence in the voice of the jester: "There is also strong experimental evidence in favor of the conjecture from the fact that we have not been invaded by hordes of tourists from the future."[12]

Hawking has used the "tourists from the future" line several times since, and it always gets a laugh. But any careful listener— and any reader of the article—would know what Hawking knew: it was *only* a joke. The question raised by the absence of tourists is answered and undercut by the paper itself, which shows that, by definition, one cannot use a time machine to return to a moment before it was created.[13] Therefore, the apparent fact that we see no tourists from the future does not mean that time travel is impossible. It means—at most—only that no time machine has been created *yet*.[14]

CHAPTER SEVEN

NOVIKOV'S WILD IDEAS

IN WHICH A PHYSICIST PROPOSES A MEANS BY WHICH
A TIME MACHINE MIGHT NOT BE INVENTED,
BUT RATHER DISCOVERED

By early spring of 1992, Hawking's chronology protection conjecture had attracted numerous supporters, several of whom had suggested other natural processes by which it might be enforced. Naturally enough, the idea had also gained its share of critics. A few of them observed that history might be repeating itself. They noted that, in the 1990s, Hawking might be said to represent the physics establishment, and they recalled that members of earlier establishments—first Eddington, then Wheeler—had conjectured the existence of a natural law that would prohibit black holes.

Igor Novikov was fully aware of all the arguments for and against the viability of time travel, and it was eminently clear to him that the question would not be resolved anytime soon. Quite the contrary, it would require, in his words, "a long period of investigations."[1] It so happened that Novikov himself was a determined optimist on the subject, and until Hawking and others produced overwhelming evidence that the laws of physics prohibited time travel, he saw no reason to suspend his own inquiries, and every reason to probe more deeply. First, though, there was a bit of tidying up to do.

The self-consistent trajectories for the time-traveling billiard ball that had been discovered by Echeverria, Klinkhammer, and Thorne had by this time been alluded to in half a dozen papers. Novikov conceded that those trajectories alone did not make an airtight case for his principle of self-consistency, as the conditions that made them possible were somewhat contrived. So he imagined versions of the experiment that removed those conditions and still yielded self-consistent results.

Support for Self-Consistency

Novikov acknowledged that the billiard ball's self-consistent histories were possible in part because the experiment allowed for slightly different trajectories into the right-hand, time-dilated mouth. It was these trajectories that made it possible for the ball to emerge from the left, non-time-dilated mouth at an angle that would cause it to strike its earlier self with only a "gentle, glancing blow," set it spinning as it traveled, and so make possible its own self-consistent history. Suppose, Novikov said, we restrict those trajectories and disallow anything like a ball's "spin." We might do so with two adjoined tubes and a piston.

Imagine that the tubes are laid along Polchinski's paradoxical trajectory. The first tube begins at a point off the centerline between the holes and runs into the right, time-dilated mouth. The second begins at the left, non-time-dilated mouth and runs to meet the first tube at a junction halfway along its length. In a paradoxical solution, a piston emerging from the left mouth and traveling inside the tube would block its younger self, thus preventing its own existence.

Even given these restrictions, Novikov could imagine a self-consistent solution. The piston emerging from the left, non-time-dilated mouth would travel at a slightly slower speed, such that it would not reach the junction in time to block the path of its younger self. But its forward surface would reach the junction opening just barely, when it would be brushed by the side of the

younger piston. The younger piston would thereby be slowed sufficiently that, when it emerged from the left, non-time-dilated mouth
as the older piston, it would be traveling at a slightly slower speed.
Its history would be self-consistent.

The self-consistent billiard ball history discovered by Echeverria
and Klinkhammer was possible for another reason: the older ball
was able to impart a "gentle, glancing blow" to its younger self.
Suppose, Novikov said, we disallow such an interaction. We build
into the ball a bomb with a hair-trigger fuse, ensuring that when it
emerges from the left mouth and strikes its younger self, the
younger self explodes into fragments. Novikov reasoned that if the
younger ball exploded before it reached the right, time-dilated
mouth, then it simply could not have emerged from the left, non-
time-dilated mouth. Rather, a fragment of the exploded ball would
have traveled into the right mouth and emerged from the left mouth
to strike the unexploded ball.

The reader may feel that such solutions are post hoc, as do
indeed a great many scientists. Post hoc reasoning is a radical
departure from a most basic tenet of scientific inquiry—namely,
that cause precedes effect. But Novikov would respond that our
antipathy evolved in an environment in which causality has one
direction. A universe that contains closed timelike curves, on the
other hand, is a universe whose future can affect its past. In such a
universe, post hoc reasoning becomes viable and even necessary.

It was through such reasoning that Novikov could argue that,
in Echeverria and Klinkhammer's solution, the emergence of the
billiard ball from the left, non-time-dilated mouth meant that it
must have entered the right, time-dilated mouth. It made no sense
to say that if the ball took a certain trajectory once, it then had a
second chance to take another trajectory. The ball could not have
one history and then another history. Polchinski's paradox, which
presupposed that the billiard ball could have two trajectories and
two worldlines, was based upon an obviously flawed assumption.

Novikov imagined still other self-consistent histories—among them a billiard ball into which is built a bomb and a radio transmitter that would signal its earlier self to explode, and a ball made of uranium just below critical mass that would collide with its earlier self and so destroy it. Interactions more complex than these were difficult, if not impossible, to model mathematically. Nonetheless, Novikov maintained that they could also be expected to be self-consistent. He had arrived at David Lewis's conclusion by another path.

A New Use for Time Machines

Even as Novikov was shoring up the principle of self-consistency, he and a physicist named Andrei Lossev were pushing the larger inquiry into new territory. In July 1991 they had sent a paper to the journal *Classical and Quantum Gravity*, and in May of the following year it was published as "The Jinn of the Time Machine: Non-trivial Self-Consistent Solutions." The title requires a bit of parsing.

Working through it backward, we know that self-consistent solutions were Novikov's broad answer to the problem of paradoxes. How might they be nontrivial? To a physicist, a trivial solution is one that is obvious and uninteresting. Consider an equation in which the sum of two unknowns equals 3. A trivial solution might have one unknown equal to 1 and the other equal to 2. It would be perfectly correct, but it would also be uninteresting and obvious, and would have required little or no real effort to prove. A solution that had one unknown equal to +103 and the other equal to −100 would be both more interesting and less expected, and in the right context it would suggest new avenues of inquiry or a fresh perspective on a familiar problem. It would be nontrivial.

The phrase "time machine" is noteworthy here because in spring of 1992 it was still rare in the professional literature; the term would, of course, be recognized by any reader. But the second word in the title would make even a theoretical physicist reach for a dictionary. *Jinn* are mentioned throughout the Koran as a race of spirits that can appear suddenly and unexpectedly. Lossev and Novikov had borrowed the word and adopted it to describe something that, most would probably agree, merited its own name. It was an entity that could be produced, quite literally, from nothing.[2]

There is at least one way in which a time machine might enable one to *seem* to create something from nothing. To take a case at hand, suppose I am an author writing against a deadline. I choose a time and place that I am reasonably certain will be conducive to writing—say, Greenwich Village in 1955. I toss my half-completed manuscript and some hundred dollar bills from that period into a travel bag and jump into a mouth of a time machine in my home, then emerge from the other mouth on Seventh Avenue South and Christopher Street during the last year of the first Eisenhower administration. I check into a nearby hotel and begin work. Some months later, with a completed manuscript, I jump back into the 1955 mouth and emerge from the other mouth in my home, say, a second after I left. This use of a closed timelike curve yields the desired result—that is, a completed manuscript well in advance of the deadline—but the method is less than ideal. Although I returned to the present only a second after I left, my own chronological age has increased by the duration I lived in 1955. Moreover, I in no way diminished the actual work: I merely shifted it to another place and time. A physicist might call this a trivial solution.

Suppose I am faced with the same problem (a need to produce a manuscript before a deadline) and given the same tool to resolve it (a time machine). By exploiting the second type of temporal paradox— the *bootstrap paradox*—I might solve the problem in a way that would be decidedly nontrivial. The result would be a manuscript that did not merely *seem* to be created from nothing; it would actually *be* created from nothing.

The Bootstrap Paradox

Imagine that my time machine has one mouth in the present and the other in the future. I jump into the "present" mouth and emerge from the "future" mouth at a place of my choosing (say, my home) at a time of my choosing (say, the day before I send a copy of the manuscript to the publisher). I walk into the kitchen, where I see that my future self has left the manuscript on the counter. I fold the ream of pages under my arm, jump back into the future mouth, and emerge from the present mouth a moment after I left, with manuscript in hand. Of course, I cannot send this manuscript to the publisher: if I did, I would not have it to place on the counter to be retrieved by my past self. So I make a copy and remind myself that on the day I send it, I must also leave the original on the counter.

We might imagine permutations. In one, it is my future self who does the time traveling, delivering the completed manuscript to me in the present. I copy the original and remind myself that sometime in the future I will need to take it into a closed timelike curve and deliver it to my (then) past self. In another, my future self is uncooperative. I must travel into the future, "borrow" the manuscript and take it back to the present for as long as I need to copy it, then return the original to the future a moment after I took it, before my future self notices that it is missing. (How I might trick my [more knowledgeable] future self is a problem addressed in several science fiction stories.) In yet another permutation, a more agreeable future self travels through the time machine to deliver the manuscript to me, waits long enough for me to make a copy, then returns to the future with the original.

For purposes of this illustration, the particulars of back and forth are unimportant. The significant feature in each is that the original manuscript is a jinn, an entity possessed of some curious qualities.

Consider the manuscript in the first of the cases described.

Exactly where did it come from? I didn't write it. I found it left for me by my future self. But he didn't write it either. At the moment he left it on the counter, it had been in his possession since he was my present self. Consider the manuscript's past. The paper on which it was written was never milled, was never wood pulp, was never part of a tree. The earliest moment in its existence as seen from the outside universe occurred when it emerged, with me, from the present mouth. Alternatively, consider its future. The same paper will never grow yellow with age, will never crumble into dust. The most future-ward moment in its existence as seen from the outside universe will occur just before it disappears, with me, into the future mouth.

The Nature of Jinn

We might best appreciate the strangeness of the jinn manuscript by visualizing its worldline. Seen through the second law of thermodynamics, any given worldline might be said to represent a temporarily organized collection of matter and energy, and the endpoint of a worldline marks the place in spacetime at which its organization ceases.

My life may be said to begin at the moment molecules arrange themselves into something called "me," and it ends (and my worldline ends) when that particular arrangement of molecules and electromagnetic bonds disintegrates. The astrophysicist George Gamow famously (and self-effacingly) titled his 1970 autobiography *My World Line*.[3] Especially in a relatively complicated part of the universe like the surface of Earth, worldlines tend to be greatly interwoven, like threads pulled together into cords that in turn are wound into bundles of rope. A given atom—say, a hydrogen atom—has a worldline that is temporarily woven into worldlines of larger collections of matter: carbon molecules, living cells, geraniums. Nonetheless, the worldline of that single hydrogen atom remains intact, and it is likely to be very long, beginning 10^{-35} second after the big bang and ending when atoms themselves disintegrate, 10^{32} years from now.

Whatever their length or internal complexity, the worldlines of all conventional, nonjinn entities begin at one point in spacetime and end at another, and their orientation on a Minkowski diagram is mostly vertical. The worldline of a jinn, on the other hand, traces a closed loop, and for this reason a jinn is truly ephemeral—unlike anything else in the universe. Thus does Novikov's coinage become doubly appropriate. Not only do jinn described in the Koran behave differently from other beings; they are *fundamentally* different. Men are made of clay, and they return to clay after their death. Jinn, on the other hand, are made of "smokeless fire," and when they disappear they leave no trace.[4]

This behavior suggests a problem. The law of conservation of energy states that the total amount of energy in a closed system remains constant. Because the universe is a closed system, the total amount of energy in the universe remains constant over time. It is impossible to annihilate anything completely: we will always be left either with energy or with particles. Likewise, it is impossible to create something from nothing. Jinn seem a clear violation of the law of conservation of energy. In fact, however, jinn do not violate it any more than do virtual electromagnetic particles. Both come into existence quickly—but they also vanish quickly—and so the cosmic ledger remains balanced.

Still, jinn seem, at least at first glance, to violate another principle: the second law of thermodynamics, the law stating that all closed systems run down, losing order and gaining entropy. Eddington once made a wry comment on its unimpeachability:

> If someone points out to you that your pet theory of the universe is in disagreement with Maxwell's equations—then so much the worse for Maxwell's equations. If it is found to be con-

tradicted by observation—well, these experimentalists do bungle things sometimes. But if your theory is found to be against the second law of thermodynamics I can give you no hope; there is nothing for it but to collapse in deepest humiliation.[5]

To understand the relation of the second law to jinn, let us return to a case at hand—my jinn manuscript. Over time the chemicals in the ink will react with both the paper and the atmosphere and fade, and the paper will likewise react with the atmosphere and age. Even during the brief interval between the moment I retrieve it from the future and the moment I return it to the present and begin to copy it, we would expect that the manuscript will decompose slightly. Moreover, we would trust that an analysis of sufficient sensitivity (say, with an electron microscope) would reveal degrees of decomposition along its brief worldline. The problem is that, if the closed time-like curve is to be self-consistent and if the jinn is to be a jinn, then the manuscript I leave on the counter must be in the same condition as the manuscript I take from the counter—*exactly* the same condition, even at the molecular level. The second law of thermodynamics would seem to make such a state of affairs impossible.

Novikov believed otherwise. First in his contribution to the consortium paper in 1990, and then with Lossev in the jinn paper, he argued that, as it traces its circle through spacetime, a jinn may "self-organize," losing entropy and gaining order. We might illustrate as follows. Imagine the worldline of the jinn manuscript as a circle of 360 degrees. At 0 degrees the manuscript is at its most futureward point; it is also at its state of greatest decomposition and greatest entropy. I take it into the future mouth of the time machine and move clockwise on the circle, emerging from the present mouth with manuscript in hand, at the 180-degree mark. During its journey from 0 degrees to 180 degrees, the manuscript has somehow reversed its decomposition, losing entropy and gaining order. Then, for the remaining half of its worldline it decomposes as might any nonjinn entity, until at 360 degrees, when it is back on the kitchen counter, it is in precisely the same condition it was in when I retrieved it.

The question of how a jinn manages such a trick is a difficult one. But Lossev and Novikov described several means, the simplest of which features two interacting billiard balls. One is a "nonjinn" ball that loses energy over time. The other—the focus of our interest—is a jinn billiard ball that, in its circuit along a closed timelike curve, must somehow regain energy. Because this is a thought experiment, we may simplify matters by allowing the jinn ball to lose and gain only *heat* energy (which we measure as temperature), requiring that its other energies remain unchanged.

Suppose the jinn ball, about to enter the right, time-dilated mouth of the time machine, is just outside that mouth. We call this its initial position. The ball is slightly warmer than its immediate environment. It enters the mouth, travels through the throat and emerges from the left, non-time-dilated mouth. As expected, the ball's temperature has dropped slightly. If its path in spacetime is to be self-consistent, the jinn ball must somehow regain the lost heat by the time it returns to its initial position. So, upon emerging from the left mouth, it collides with a nonjinn billiard ball (not an earlier self this time, but an entirely different ball).

Because we have required all the jinn ball's energies except heat energy to remain unchanged, the jinn ball cannot lose kinetic energy—that is, velocity or the energy of motion. But we have made no restrictions on the energies of the nonjinn ball. It *can* lose velocity, and it does. During the collision, some of the kinetic energy of the nonjinn ball is transferred to the jinn ball in the form of heat, and the temperature of the jinn ball rises to a level slightly above what we recorded initially. Meanwhile, the collision has sent it along a path that takes it through its initial position, and at the moment it reaches that position, its temperature has fallen to exactly the temperature it had initially.

What has happened? Because we allowed the jinn ball's heat energy to vary and required that its other energies remain constant, we "forced" it to draw upon an external system to return to its initial state.

Imagining Jinn

Let us examine another illustration of jinn—one with a bit more texture. Escher's 1961 lithograph *Waterfall* (Figure 7.1) borrowed and embellished another of Roger Penrose's impossible forms—the "tribar" shown in Figure 5.2a. Escher regarded his elaboration of the tribar as he regarded all his work—as interesting and provocative, no more. Although he was aware of Einstein's theories of special and general relativity, and in fact completed several works inspired by them, he did not connect *Waterfall* to closed timelike curves. Nonetheless, one cannot help but observe that the image is strange in several ways, and they are precisely the ways in which closed timelike curves are strange.

It is impossible to find the place where the water begins. Choose any part of the channel, and all seems normal. If we imagine the point of view of a leaf floating on the surface of the water, we may follow the stream through its entire course. As with the procession of monks in *Ascending and Descending*, there is no place along the channel in *Waterfall* at which we sense an interruption, no point at which we would choose above any other as a beginning or an ending. Like the procession of monks, the channel is utterly self-consistent. Only when we regard it from outside—from the point of view of, say, the man leaning against the courtyard wall or the woman hanging laundry—do we sense that something is odd. As in *Ascending and Descending*, the feature of *Waterfall* that is intriguing (or disturbing, if one is in a certain mind-set) is that it contains two reference frames that, intuition and experience tell us, ought not coexist. One is that of the water; the other is that of everything

Figure 7.1 *Waterfall*

else. Although Escher nowhere encouraged such an interpretation, perhaps the only way to make sense of the image is to regard the water as a jinn. The water in Escher's image has the same relation to the world outside it as a jinn worldline has to the spacetime in which it is embedded.

Ordinary water in an ordinary channel loses order and gains entropy over time, and a given volume downstream from any other given volume will have less order and more entropy. Its potential energy—that is, the energy it possesses by virtue of being in a gravitational field—has become kinetic energy. Yet studying Escher's image for a moment reveals that the water, as it courses from the base of the waterfall along the zigzagging channel to the point just above the ledge, *gains* order and potential energy. As it cascades over the ledge, it suddenly and rapidly loses the order it has gained, and its potential energy becomes kinetic energy until, at the base of the waterfall, it begins to gain order and potential energy again. (Whereas the worldline of the jinn manuscript moved clockwise and located its greatest order at 180 degrees, the water jinn moves counterclockwise and locates its greatest order at 0 degrees.)

The water in Escher's image, like the jinn manuscript and the jinn billiard ball, interacts with systems outside itself. It moves the waterwheel and, we might reasonably suppose, makes possible the surrounding life and activity—the plants, the people, the wash on the clothesline. Moreover, if we are to give credence to a wryly practical comment in Escher's lecture notes, the water can be affected by systems outside it. "The miller," the notes say, "can keep [the current] perpetually moving by adding every now and then a bucket of water to check the evaporation."[6] Might this be the means by which it gains order? Realistically, a bucket of water delivered at widely spaced intervals is insufficient to the cause, but if the water is poured with enough force it might, at least momentarily, reverse the direction of a small part of the current. At least minimally then, our water jinn does draw upon an outside system for energy and raw materials.

These examples—water's managing to coerce order and energy from a man with a bucket, or a billiard ball's managing to gain heat from a collision with another ball—may strike us as unrealistic and contrived. But we should recall that the universe we know contains entities that habitually draw energy and material from outside themselves, thereby self-organizing and gaining order. These are, of course, living organisms. An acorn draws energy in the form of sunlight, and raw material in the form of water, carbon dioxide, and organic matter; over time it "reorders" these into an oak tree. Lossev and Novikov suggested that black holes, a class of objects whose internal organization is far simpler, are also capable of self-organization. They attract matter and energy, thereby gaining mass and, if conditions are right, angular momentum. In short, the jinn's capacity for self-organization has precedent and parallels with which we are familiar. That a jinn by definition must self-organize, then, is not in and of itself reason to dismiss it as unrealistic.

Let us extend the parallel further. One can assist and even make possible the self-organization of an acorn by pushing it into moist and fertile soil. One could assist the self-organization of a black hole simply by dropping mass into it. Could we create conditions conducive to the self-organization of a jinn? In the first of the preceding examples, such assistance would be a simple matter of a well-placed billiard ball. In the second it would be rather more difficult but might be managed with many buckets of water directed up-channel with great force. Reversing the decomposition of my manuscript would be a considerably greater challenge but, given a small army of manuscript preservationists and restorers, not impossible. Their work, of course, would be to get the manuscript to a condition somewhat better than it was in when I received it, so that it might be allowed to degrade to that (initial) condition.

We now have something of a better understanding of jinn, and the jinn manuscript may for some reasons seem *less* probable than we first suspected, and in some ways *more* probable. Let us reconsider the conditions necessary to its creation. I would need to be certain that my future self possessed a completed manuscript and that he was willing to allow me to retrieve it, or willing to deliver it to me. I would need to be certain of the existence of systems along the manuscript's worldline from which it could draw energy and matter to self-organize. And, of course, I would have to be certain that my future self had access to a time machine before the submission deadline. Of the three conditions, the last is the one on which the other two depend, and it may well be the most difficult to meet.

In 1992, astrophysicist Neal Katz half seriously proposed a means by which one might ascertain the viability of time travel. It was utterly straightforward and disarmingly low-tech: Publish an article in a professional journal that is subscribed to by libraries worldwide and so has a good chance of surviving in multiple copies into a reasonably distant future. The article would include the following request, printed in boldface capital letters: "IF TIME TRAVEL IS POSSIBLE AT ANY FUTURE DATE, PLEASE CONTACT THE AUTHORS ON [PUBLICATION DATE], 1992." It would be immediately followed by the author's address.

In the spirit of thought experiments, let us examine the proposal seriously. Of course, the time machines described already, as well as those that will be introduced later, would allow travel pastward only to the moment of their construction. If Katz is to expect a knock on his door, he must presuppose that a time machine exists in his own present. He must also presuppose not only that his archived message will be received by someone with access to a time machine, but that that person will act upon his request. As long as libraries endure, any archived journal has a fair chance of surviving,[7] but what of the presupposition that the recipient is cooperative? This seems somewhat less reasonable. In Shakespeare's *Henry IV, Part 1*, when Glendower says, "I can call spirits from the vasty deep," Hotspur deflates the boast with a commonsense response: "Why, so can I, or so can any man; But will they come when you do call for them?" Indeed, like Glendower's spirits, a time traveler might prefer not to appear.[8]

The reader worrying that we are taking speculation too far from terra firma may be somewhat comforted to learn that we are not alone. In "The Jinn of the Time Machine," Lossev and Novikov not only considered these problems; they also devised a solution for them. The first two-thirds of the paper suggested means by which jinn might self-organize on their circuits through spacetime. Such concerns were rather removed from mainstream theoretical physics, and their presentation was no mean feat, in that, as we have seen, theoretical physics defines its mainstream rather liberally. Still, the authors were only warming up to their real subject—a possibility that, in terms of strangeness, was quite unlike anything in the professional literature.

The Two Kinds of Jinn

As intellectual property attorneys remind us, a printed or electronically stored work has two components: the paper and ink or computer disks, and their content. The separability of these components allowed Lossev and Novikov to define two kinds of jinn. A jinn made of ordinary atoms is what the authors termed "jinn of the first kind." Their paper also described a nonmaterial jinn—one made purely of information. This, naturally enough, they called a "jinn of the second kind."

We might generate a jinn of the second kind by modifying the worldline of the jinn manuscript. Suppose, as before, that my future self completes the manuscript on the day before he sends it to the publisher. But I do not travel into the future to retrieve it; nor does he travel into the past to deliver it. Rather, he scans it onto a computer drive, whereupon it is transmitted via radio waves into the time machine's future mouth. The radio waves emerge from the present mouth, are intercepted by my receiver, downloaded onto my computer drive as the tiny magnetized regions called bits, and translated back into signals that my computer software translates back into the original alphanumeric symbols of the print manuscript. I remind myself that, on the day before I send the manuscript to the publisher, I must transmit its contents to my (then) past self.

The worldline of this manuscript is traced by at least four identifiable forms: paper and ink, electronic signals, bits, and radio waves. When we talk about the manuscript, though, we are talking about none of these things. What we are talking about is pure information. Yet it is no less susceptible to the second law. As anyone who has played the children's game "telephone" will attest, information will degrade over time, losing order and gaining entropy. Given time, some of the manuscript, perhaps as small as part of a cross on a t or a tail on a y will go missing—a casualty of imperfect transmission and entropy.

Lossev and Norikov's larger point, though, was that a jinn of the second kind might represent knowledge created from nothing. Of course, such knowledge could be generated only through a time machine. It was possible, they surmised, that access to one might be easier than many had assumed. They imagined a scheme in which, by exploiting the circular logic of the principle of self-consistency, a jinn of the second kind would enable the discovery of a time machine that would, in turn, ensure the jinn's creation.

Natural Time Machines

In the four years since the Morris-Thorne-Yurtsever paper, the possibility of *preexisting* time machines had been tossed about. These would be wormholes evolved into time machines as, after the manner of Frolov and Novikov's 1990 paper, their mouths shifted reference frames. For his part, Thorne doubted the existence of natural wormholes, suggesting that it was difficult to imagine how two singularities could somehow find each other in hyperspace and form a throat.[9] Moreover, although the laws of physics seem, at least in hindsight, to have made phenomena like black holes inevitable, there seemed to be no reason to expect that the universe, left to its own devices, would ever evolve a wormhole. Nonetheless, some admitted it was possible that wormholes existed in the spacetime singularity that preceded the big bang, and survived into our time.

The reasoning was that during the Planck epoch, that first sliver of time in our universe's life, all spacetime was quantum foam. Its complex and probabilistic topography almost certainly included wormholes, and it was possible that some of them had been inflated to macroscopic size during the inflationary epoch, and live somewhere in the universe today. There is no observational evidence that such is the case; most astronomers expect that if our universe were threaded with wormholes, the overall distribution of matter would not be nearly as smooth as it is.[10] Still, as far as many physicists are concerned, this does not mean that there were no wormholes during the Planck epoch; neither does it mean that none survived inflation. Why, then, do we see no traces?

It may be that if, as Frolov and Novikov suggested, a wormhole—once created—almost inevitably begins to form a closed timelike curve; if Hawking's intuitions are correct, then before the process gets very far, an undefined chronology protection mechanism kicks in and destroys it. In other words, there *are* wormholes, but they are extremely short-lived. Or perhaps there are types of wormholes, and only one type survived inflation, making for a very sparse population. Plenty are out there, but the universe is very, very big, and most are far beyond the part of the universe we can see.

In "The Jinn of the Time Machine," Lossev and Novikov did not ask their readers to suppose that wormhole time machines would exist at some moment in the future. Instead, they asked them to suppose that they do exist (or at least that *one* exists) somewhere in the universe in the present. Suppose, they said, that somewhere in space, at a distance reachable by a spacecraft, there is a wormhole time machine. Perhaps it was constructed by that hypothetical supercivilization and then abandoned; perhaps it outlived them; or perhaps it was natural—a remnant from the early universe. For the authors' purposes its origin mattered no more than did its precise distance from us. If it simply exists, they said, we could use it, and no technology indistinguishable from magic would be necessary. In fact, it could be reached—and *used*—with technology that we possess now or may possess shortly.

The Clever Spacecraft

To accomplish time travel, the authors claimed, we would need access to only three components: sufficient raw materials to construct a spacecraft capable of a very long voyage, an advanced computer that could download and replicate the spacecraft design, and an automated plant that could process the materials and construct the spacecraft *from* that design. Perhaps most interestingly, as a means of ensuring that our program is carried out, we would also need to guarantee our own absence from it. The paper says that we assemble materials, construct such a computer and such a plant, and simply get out of the way. At this point a dramatic consequence of the closed timelike curve is manifest. Immediately, they say, a very old and rather battered spacecraft lands. Then, with no muss or fuss, it downloads its own design and the location of the mouths of the wormhole time machine into the advanced computer. Its work is finished.

The advanced computer then instructs the automated plant to construct a new spacecraft based on the design of the old spacecraft, and identical to the old spacecraft. When the new spacecraft is complete, the computer downloads the location of the mouths into its navigation system. The new spacecraft lifts off, achieves orbit, confirms the health of its systems, and readies itself for the long voyage ahead. It fires its engines again and launches itself out of orbit, then out of the Solar System in the direction of the present mouth. Once in interstellar space, it shuts down most systems and simply coasts.

Engineers have already imagined means by which a spacecraft might travel at 10 or 12 percent of the speed of light. But even at such velocities, it would take centuries to reach many of the nearer stars. Consequently, although the technology required for unmanned interstellar travel would be decidedly twenty-first-century, the social and cultural commitment would be, for lack of a better word, medieval. Like the designers of the cathedral at Chartres, the planners of a given interstellar mission could not expect to see its fulfillment, and would regard themselves as a small part of an endeavor that would be concluded only long after their deaths.

By contrast, Lossev and Novikov's clever spacecraft would ask of its designers no such patience. The results of a successful mission would appear a moment after the advanced computers and the automated plant were complete. From the point of view of the spacecraft, however, the duration of the mission would be no shorter than those of more conventional interstellar probes, and whatever its means of propulsion, the spacecraft would need to survive a very long time. This might present the project's greatest challenge. The interstellar medium—a hydrogen atom in every cubic centimeter—would gradually erode any material moving through it at an appreciable speed. Microscopic grains of interstellar dust are far less common (there is perhaps one grain for every trillion or so atoms), but at high speeds a single impact could cause explosive damage. Internal malfunctions would pose at least as great a danger.

In the 1970s, such were the considerations of a group of scientists and engineers who called themselves the British Interplanetary Society, and who made a design study for an unmanned interstellar mission. Coasting at 0.167 the speed of light, *Daedalus* would have made a one-way flyby of Barnard's Star, chosen as a target because at six light-years' distance it was relatively nearby, and because its "wobbling" had suggested it was alternately pulled and pushed by a planet in its orbit. *Daedalus* would have released smaller probes in the star's vicinity.

The spacecraft structure, communications system, and much of its payload could have been built with twentieth-century technology. It was to be propelled by "pulsed fusion," a sequence of small nuclear explosives discharged and then ignited behind it. It would have been protected from the interstellar medium and dust grains by a beryllium shield thirty-two meters across, and safeguarded against internal malfunctions by a simple design, a computer fully capable of identifying and diagnosing problems, and robots roaming inside the spacecraft performing routine maintenance and repairs.

Some thirty years after its conception, Project *Daedalus* remains the most detailed and comprehensive study for an interstellar probe, but it is far from the last word. Although our actual experience with such spacecraft is meager,[11] those working in the more theoretical end of aerospace engineering have produced a surprisingly large body of research. During the 1970s and 1980s, NASA examined concepts for two unmanned missions to space beyond the Solar System, and in 1999 the agency formed the Interstellar Probe Science and Technology Definition Team to come up with still other ideas.

So let us imagine the clever spacecraft as a next-generation *Daedalus*. Although dust grains are far more rare in deep space than within any star system, even one strike every thousand years takes a toll. A microscopic examination would reveal a surface pitted and scarred. Likewise, once in a great while a gamma ray strikes and short-circuits an electrical component. But the spacecraft's most important systems have several backups, and all its systems are well shielded.

As it completes the first leg of its journey and enters the vicinity of the present mouth of the wormhole, the spacecraft is still performing reasonably well. It fires braking rockets and slows. Telescopic cameras aimed in the direction of the mouth discern something that looks like a mirrored sphere.[12] Soon it becomes apparent that it is not a mirror at all. The stars visible on its surface are not reflections of the stars in this part of space. They are stars from somewhere else. And they are not "on the surface." Rather, they are in the part of space that is visible *through* the sphere. The spacecraft takes a last navigational fix on the constellations and passes through the surface. Inside the throat, its cameras, radar, and other remote sensors probe the walls along which it is passing, which Sagan had called "deeply mysterious."[13] Almost immediately it exits from the past mouth. Rear-facing cameras see a sphere full of stars—and again, they are not the stars in this part of the sky, but stars as seen from the vicinity of the present mouth.

How does the spacecraft know where in the universe it is? An ordinary spacecraft might search for identifiable radio sources outside the galaxy, then stars in this galaxy with known spectral lines, and use them to triangulate its position. But the clever spacecraft needs to make no such effort, because it already carries such information with it.

There is a second navigation challenge, owing to the fact that the spacecraft has traveled pastward. How does it know *when* it is? An ordinary spacecraft might determine the date by appealing to various natural cosmological clocks—like the slowing rotation of pulsars. But once again, the clever spacecraft carries such information with it. It already knows the date. It also knows how long it will take to return to Earth. If it can complete its homeward journey in time to coincide with its departure, then it begins immediately. If, on the other hand, it is so far from Earth that it cannot complete its return in that time, it has a fail-safe procedure. Much like the billiard ball that Thorne imagined in his meditative moment, the spacecraft can, simply by returning to the other mouth and using the time machine over and over, travel further and further into the past, and so buy for itself all the time it needs.[14] In either case it eventually returns to Earth as the "very old and rather battered" spacecraft described earlier.

The more time the clever spacecraft spends traveling through space, the greater the chance it will be hit by a micrometeoroid or a gamma ray, or suffer a hardware malfunction. For these reasons, decreasing total travel time might be a good idea. If our purpose is to convey information—specifically, the spacecraft design and the locations of the wormhole mouths—then the spacecraft itself need not make the whole journey. Instead, on its emergence from the "past" mouth it could simply send a signal to a receiver on Earth.

In either case, the principle of self-consistency has been used to generate, quite literally from nothing, both a spacecraft design and knowledge of the locations of wormhole mouths, and employ them to discover—and to use—a time machine. If Lossev and Novikov are right, then we may not have to wait for that suitably advanced civilization; *this* civilization could use a time machine—if not immediately, then in the very near future.

Such a scheme seems at least possible. But for reasons that we may find difficult to pinpoint, it strikes us as problematic. Our reservations may have less to do with spacecraft hardware or longevity than they do with rather more philosophical issues. We may be disturbed by the scheme's supposition of a predetermined future. (Clearly, such a prospect did not bother Lossev and Novikov; indeed, what they did with the clever spacecraft was to use that future to advantage.)[15] Or we might be troubled by the scheme's "getting something from nothing" aspect. (This objection, shared by several physicists, will be discussed in Chapter 10.) That the clever spacecraft was given a hearing at all was some measure of how broad the discussion had become.

By early 1992, all these ideas—the putative mechanisms of chronology protection, the difficulties that vacuum fluctuations present to time travel, ideas of self-consistency and jinn—had appeared in respected journals, and the inquiry itself had developed arguments and counterarguments. Still, most of the interchange among the authors had been indirect. It was clear that more immediate interaction would be in everyone's interest. It seemed the right time for a face-to-face discussion.

CHAPTER EIGHT

A TIME TRAVEL SUMMIT

The university-based institute—that is, a community dedicating itself to research and academic discourse—is a European invention. Many flourished through the nineteenth century, and most had a well-defined disciplinary focus and a general air of exclusivity. Such features made them difficult to transplant to the United States, where scientific practice was less rigidly structured, and university campuses, especially by comparison, were fairly democratic. But the American model had come at a cost: by the first years of the twentieth century, scientific progress in the United States was lagging behind its European counterpart. To many, the reason was obvious: American researchers had no place for sustained and uninterrupted work.

In time, American iterations—the Carnegie Institution of Washington, the Institute for Advanced Study in Princeton, and the Nobel institutes for physics—would become as important to scientific progress as their Old World forebears. Their histories make fascinating reading, and most involve twinned figures: a person of means with a need to contribute to society but lacking a focused vision, and an academic sort, in mid or late career, possessed of a vision but with no means to realize it.

As regards the second figure, one American institute is an interesting exception, owing its existence to a twenty-seven-year-old graduate student. His name was George Stranahan, and in 1957 he was working through the last stages of a PhD in physics at the Carnegie Institute of Technology in Pittsburgh. Stranahan was the fortunate son of a prosperous family—they had founded the Champion Spark Plug Company—and when he was writing his dissertation he could afford to rent a house in Aspen, Colorado. The views there were breathtaking, and the trout fishing was unmatched. Yet his work suffered, and he knew the reason. He was lacking the company of other physicists and the intellectual stimulation that came with it.

As it happened, the area was already home to an institute—a well-known one. The Aspen Institute for Humanistic Studies had been founded in 1950 and had since gained international acclaim as a meeting place for scholars, as well as political and corporate leaders. Stranahan contacted the director, Robert W. Craig, and Michael Cohen, a University of Pennsylvania physicist who he knew was like-minded. Together, the three drew up a proposal for a physics community dedicated to face-to-face interaction, and they presented it to the institute's CEO, a man named Robert Anderson.

Anderson knew that the institute's original mission statement had called for a synthesis of the humanities and sciences, and that the second aspect had been neglected. A physics division within the existing institution would fulfill that mandate and then some. So he set aside four acres of a part of the campus called Aspen Meadows, near Aspen's historic West End. Stranahan contributed some family money and persuaded several universities to lend assistance. In the summer of 1962, the first building was opened, and forty-five physicists from twenty-two institutions around the world arrived for several weeks of seminars and informal discussions. It was an auspicious beginning for the institute's newest division.

The conferences attended by professional physicists are, as a rule, rigorously structured affairs. Presenters at the annual meeting of the American Physical Society are allowed twelve minutes to speak and two minutes for questions, and at any moment during the conference as many as ten presentations might be going on simultaneously. Even smaller and more specialized conferences usually mean two or three days crammed with twenty-minute presentations. Less formalized give-and-take and actual conversation, if it happens at all, is squeezed into a few minutes during lunch and coffee breaks, and a participant might come away more exhausted than invigorated, with a few scribbled e-mail addresses and a vague promise from someone to talk more later. Things were a bit more relaxed in the 1960s, yet most felt there was still too little time for sustained interaction. If there was one thing Aspen could provide, Stranahan thought, it was time. His hope was that physicists would start collaborations and actually begin work there and then.

In the past two decades that hope has been realized several times over. The broad outlines of the Fermi National Accelerator Laboratory (Fermilab) were drawn up in Aspen, and it was there that *string theory*—perhaps the most promising path to understanding quantum gravity—saw its earliest formulations.

In 1968 the physics division was able to sever its legal affiliation with the institute, to become an independent nonprofit called the Aspen Center for Physics. By the 1990s, theoretical physicists regarded the center as a sort of nirvana, and had come to call it simply "Aspen." So many had taken up permanent residence in the environs that the local yellow pages actually listed a section for physicists. Most anyone in the field would expect to spend at least some time at the institute—some for their own research, most to attend conferences or relatively informal gatherings called workshops. Skim through the acknowledgments of articles in *Physics Review*, and you will see many thanks for summers at Aspen.

Since its founding, the Aspen Institute has accumulated a rich history of brilliant minds at work and play. There have been moments of comedy. During one evening gathering, Nobel Laureate Murray Gell-Mann played the part of a "crackpot" cosmologist, erupting from the audience and jumping to the stage to expound upon a "theory of theories" until two accomplices pulled him off. There have been moments of profound sadness. Physicist Heinz Pagels was the executive director of the New York Academy of Sciences and a talented author who happened to be married to religious scholar Elaine Pagels. In many ways he embodied the institute's synthesizing vision. Pagels was an avid technical rock climber, and the surrounding mountains offered him many opportunities. In 1988, while scaling nearby Pyramid Peak, he fell to his death. On balance, though, the experiences of Aspen were of renewal. The institute's music school shared the larger campus, and the orchestra practiced in an open-air music tent. Stephen Hawking once recalled that it was there that he first heard Francis Poulenc's *Gloria*.

Until the early 1990s, most Aspen conferences had focused on areas like particle physics or wave mechanics. But in 1991, James York, a relativist at the University of North Carolina, organized a conference on mathematical relativity. It turned out to be a stimulating two weeks, and one of those attending—a relativist named David Garfinkle—was inspired enough to consider proposing a workshop on another subject of interest to relativists: "cosmic censorship."

Some thirty years earlier, Roger Penrose had sought to discover a set of conditions by which a stellar implosion might produce a black hole without an event horizon—a *naked singularity*. Although he was unable to identify those conditions, he was also unable to prove them impossible. But a singularity sitting there in the open for all to see struck him intuitively as wrong, so he proposed the *cosmic censorship conjecture*—a proposition that a set of undiscovered laws require all singularities to exist beneath an event horizon, to be "clothed."[1]

It had been more than two decades since Penrose first made the conjecture. Despite a great deal of work in the area, no one had proved Penrose right, and no one had proved him wrong either. The question was far from settled.

Garfinkle thought an extended discussion on cosmic censorship was long overdue. He ran his idea past York, who agreed. But as interesting as the subject was, York noted that the workshop selection committee would find it more attractive if Garfinkle could couple it with another, more current issue. He suggested chronology protection. Since the Morris-Thorne-Yurtsever "time machines" letter in September 1988, there had been a number of articles on the subject, and they had received considerable attention. It was easy to see how chronology protection might be related to cosmic censorship—or, framed in more positive terms, how closed timelike curves might be related to naked singularities. Both were hypothetical phenomena at the edge of known physics, and both were enabling scientists to tease out new understandings of the universe.

Garfinkle proposed a two-week workshop. In early fall of 1991, the center accepted the proposal, and in the weeks following, announcements appeared in journals and newsletters. By the spring of 1992, Garfinkle had a list of nearly thirty acceptances. Most of the consortium was represented—Novikov, Friedman, Yurtsever, and, of course, Thorne. Gott had signed on too, as had several physicists, like Thomas Roman and Deborah Konkowski, who had been drawn into the subjects more recently. Perhaps the only palpable absences were Penrose and Hawking. But by the time the workshop convened, their thoughts were very much in the air.

Aspen, Summer 1992

The Aspen Center for Physics had three buildings, and their designers had been careful to respect Stranahan's ideals of simplicity. By the standards of a place like the Institute for Advanced Study, accommodations were spartan. Visiting researchers who did not rent apartments in town lived in modestly furnished rooms; they worked in phoneless, shared offices and had access to only a few desktop computers. All this thrift was by design. Stranahan once remarked, half jokingly, that the number of bathrooms was deliberately kept small so that a physicist answering nature's summons might need to wait behind another physicist, and thus the two would be granted an opportunity for a discussion that might otherwise never occur.

At the center of the grounds was a courtyard partly enclosed by an old grape arbor, and for presentations some of the physicists would borrow a chalkboard from a seminar room, wheel it to one end of the patio, and set up folding chairs in front of it. It was in the courtyard that, on the first day of the 1992 workshop, Thorne delivered the keynote address.

By this time, the professional repercussions of Thorne's interests that had concerned Richard Price had, in a small way, come to pass. The astronomy and physics divisions of the National Science Foundation had refused to continue to fund his research on closed timelike curves[2]—not that it mattered greatly to Thorne's own agenda, as his work had by this time moved rather vigorously into gravitational waves. In some sense his address marked the passing of a torch. He spoke on the state of research in cosmic censorship and chronology protection, and unresolved problems in both areas. In the latter case, this meant the question of whether the laws of physics prevented closed timelike curves, what protection mechanism might be involved, and what insights into quantum gravity such questions might offer.

As for the time remaining, Garfinkle was careful to keep the schedule light. But he was familiar enough with his colleagues to know that, with no framework whatsoever, they would vanish into rooms or apartments and work on projects that they had brought with them. So he had arranged several presentations. Novikov explained jinn, Roman talked about negative energy, and Gott described time travel with cosmic strings. The rest of the discourse over those two weeks—most of it, in fact—was informal, and conversations wandered far afield of anything likely to see print. One afternoon Roman, Novikov, and Thorne sat under a shade tree and speculated at length about free will and quantum indeterminacy. Naturally, there were hikes on nearby trails and evening gatherings. That summer also saw the continuance of a tradition that had been established in earlier years at Aspen: potluck dinners. The quality of the physicists' cuisine in general was better than any single physicist expected, but if no one was in a mood to cook, a large group might convene for a late repast at Little Annie's Eating House, much beloved for its all-you-can-eat ribs. Later, a few might continue the conversation at one of the city's microbreweries, where anyone sitting at the bar might overhear a discussion of vacuum stress energy or Lorentzian manifolds.

Visser

In the summer of 1992, one voice with a distinctly New Zealand accent carried a bit farther than the rest. It belonged to a theoretical physicist named Matt Visser. Visser was built like a small bear. He had a wide face, a broad nose, and the look of a man equally ready to down a few pints or join a rugby scrum, whichever came his way first. On one of the potluck evenings he showed up with two five-gallon pots of chili. Although he had spared the jalapeño for neither, one was noticeably hotter. Garfinkle sampled each, and in an inspired application of the cosmic to the culinary, pronounced the milder one "GUT scale," in reference to the energy scale during the grand unified epoch, when temperatures reached 10^{32} Kelvin. The other he called "Planck scale," in reference to the still higher temperatures of the Planck epoch.

Visser has what might be called an aggressive approach to academic discourse. He is boisterous, fearless, and usually convinced he is right. He often is. As a researcher, Visser has been astonishingly prolific. When most theoretical physicists have managed to publish a few articles in a year, he has been known to produce one per month. He took on time machines fairly early in the game, with a 1989 article that picked up where Morris and Thorne's teaching paper had left off.

A Taxonomy of Time Machines

Morris and Thorne had designed their wormhole with spherical mouths. The configuration spread outward pressures evenly and so made their calculations easier, but it also meant that anyone or anything entering either mouth would have to pass through exotic matter. For all anyone knew, exotic matter, like gravitational waves or neutrinos, would couple with atoms and molecules weakly or not at all. Nonetheless, Morris and Thorne had admitted, "It could be extremely uncomfortable for a human traveler to interact with a material that has tensions [so] large."[3] They suggested that a vacuum tube passed through the mouth might be kept clear of exotic matter and thus offer safe passage. They also noted that the problem might be overcome completely if the mouths were designed in another shape.

It was this last observation that Visser developed; his 1989 article presented a whole taxonomy of shapes. The most straightforward was a six-sided cube, with exotic matter confined to the edges and corners where the faces touched. The faces themselves were free of exotic matter and tidal forces, and anyone venturing through one of them would, in Visser's words, "simply be shunted into another universe."[4] (Although the paper does not say as much, the design might profitably be imagined as the realization of a tesseract. Anyone looking into a face would see much farther than the diameter of the cube as measured from the outside universe, and anyone who had passed through a face would be traveling at right angles to all the faces.)

There was every reason to believe, Visser continued, that more complex shapes were possible: because the length of the sides made no difference to the solution, a given face could be made any kind of rectangle or trapezoid.[5] The angles of the edges and corners, likewise, could be set arbitrarily.

Growing a Wormhole

In their 1988 teaching paper, Morris and Thorne had suggested that a suitably advanced civilization might pluck a wormhole from the quantum foam and enlarge it. It was a provocative idea, but it stopped short of saying exactly how that enlargement might be accomplished. The subsequent piece coauthored with Yurtsever had suggested quantum tunneling as a mechanism, but it was likewise vague on details. Others, like Thomas Roman, had begun to give the matter serious thought; at Aspen, in fact, he began a paper that addressed the question specifically—that is, exactly how one might grow a wormhole. On the scale of outrageous ideas, it certainly ranked, but nowhere near as high as the idea it sprang from: the "universe-in-a-lab."

Recall that, in the early 1960s, Wheeler, Thorne, and others were wondering whether a black hole could "push through" space, sever itself from it, and exist momentarily as a self-contained universe. In the late 1980s, Alan Guth, author of the original inflation hypothesis, and his MIT colleague Edward H. Farhi began to consider a similar possibility—namely, if one could reproduce the inflation of the early universe, one might create a universe artificially. It would not be momentary, but lasting. Specifically, they hypothesized that if the grand unified theories were correct, then one might induce inflation with a submicroscopic sphere of false vacuum 10^{-26} centimeter across. It would weigh only a few grams, but it would be compressed to the rather phenomenal density of 10^{80} grams per cubic centimeter—a concentration equivalent to the mass of the matter in the observable universe squeezed into an atom. Such an engineering feat would require a knowledge of high-energy physics far beyond current understanding, yet Guth and Farhi imagined two ways to do it, at least in principle.

One might heat a region of space to a very high temperature (10^{29} K) and then stop heating it, thereby allowing it to "supercool" into the false vacuum, as the early universe may have done. Or, because the inflation field is expected to interact with ordinary matter, one might compress matter to an extreme density, thus using it to push the inflation field into a bubble of false vacuum. One would then remove the matter—a neat and rather difficult trick, like pulling a tablecloth from beneath table settings. In either case, what would be left would be a false vacuum.

This false vacuum, like that which drove the universe's inflation, would expand. If it expanded too slowly, it would collapse into a black hole, and indeed Guth and Farhi expected that this is what would happen most of the time. But if it were expanding quickly enough, it would create a declivity in space, and as it grew it would stretch that declivity into a throat, where it could grow freely without being impeded by the true vacuum. In fact, by this time the true vacuum from the "parent" universe would actually assist the cause, as some would have followed through the throat. Now the bubble would be turned inside out, and the pressure would be directed outward. The bubble would disconnect quickly, 10^{-37} second after its formation.

In the summer of 1992, Roman's thinking was that a suitably advanced civilization might take a submicroscopic wormhole and embed it in a false vacuum bubble. As the space in the bubble expanded, the wormhole would grow along with it.[6] Roman was able to calculate the wormhole's properties in considerable detail: length, stresses in the throat, and so forth. But he noted that such an engineering project would face challenges that might prove insurmountable. When the false vacuum decays, much as it is thought to have done after driving the universe's inflation, its positive energy will be converted into radiation, and that radiation might flood the wormhole and trigger gravitational collapse.[7] No sooner would the wormhole inflate than it would destroy itself.

Wormhole Maintenance

So although the false vacuum might inflate the wormhole, it could not be expected to hold it open. How exactly might *that* be accomplished? It was clear that some negative energy was necessary. In his 1992 chronology protection paper, Hawking argued that any time machine constructed in a finite region of spacetime would require negative energy, and this result was independent of whether chronology protection holds. In their 1988 teaching paper, Morris and Thorne had suggested threading the mouths with *exotic matter*—that is, material with a negative energy density that would keep the walls of the mouth from collapsing inward by generating a repulsive gravitational field. They acknowledged the reservations of colleagues who suspected that negative energy might prove impossible to produce in the amounts necessary, but they preferred to take the long view, noting that, as recently as the late 1970s, physicists worked under the misconception that negative energy could not exist even on microscopic scales.

The Casimir Effect

Negative energy, strange as it may seem, is quite real and has been indirectly detected in laboratory experiments, the most famous of which involves what is called the Casimir effect, after Dutch physicist Hendrik Casimir. As physicists understand it, all space is seething with activity and, hence, energy. Yet this energy is difficult to use. To be usable, energy must be distributed unevenly. A hydroelectric dam, for instance, can operate only when water falls from a higher to a lower level. Because vacuum fluctuations are the same everywhere, the energy of the vacuum is very difficult to detect, let alone harness. But Casimir conceived of a very simple experiment that would do the job. A given region of space contains electromagnetic waves of all possible wavelengths. Casimir suggested that two uncharged parallel metal plates, positioned such that their surfaces were approximately a billionth of a meter apart, would leave between them a gap so small that only shorter wavelengths would fit within it, and longer wavelengths would be left outside. In effect, the plates reduce the fluctuations in the space between them, allowing for the generation of negative energy and pressure, which pull the plates together. The smaller the gap is, the more negative are the energy and pressure, and the stronger is the attractive force between the plates.

There are other examples of negative energy, one of which was discovered almost by accident.

Hawking Radiation

In the 1960s, Zel'dovich's research team conjectured that all the distinguishing features of a star would disappear when it collapsed into a black hole. As was mentioned in Chapter 4, Richard Price later confirmed this idea. An irregularly shaped star would implode to become a perfectly spherical black hole, and a star with a magnetic field would implode to become a black hole without one. As a star shrank below the event horizon, the irregularities and magnetic field would be shaken off and radiated away, and the result would betray no trace of the star it had been. The term used to describe such traces was "hair." (Like "black hole," the term was Wheeler's coinage.)

Black holes, it seemed, had no hair—or almost none. In fact there remained some small evidence of the star, but it was meager and limited to three very simple features: its gravitational pull, the whirlwind-like "frame dragging" discussed in Chapter 3, and its outward-directed electrical fields. Physicists had long presumed that nothing—no particle, no radiation, nothing—could ever escape a black hole. Because no signal could pass through the event horizon, there was no way to know what was happening inside it.

Then, one day in the early 1970s, John Wheeler made a small joke to a graduate student. He said that he felt a bit guilty leaving a cup of hot tea near a glass of iced tea, because the act contributed to the universe's entropy. What he meant was that, initially, the system that the hot tea and cold tea compose may be said to possess "order" in that the majority of water molecules in the hot tea are moving more rapidly than are the majority of the water molecules in the iced tea. If the cup of hot tea is allowed to cool, and the glass of iced tea is allowed to warm, eventually both will reach the same temperature, the system they compose will have become more disordered, and the universe's entropy will have increased very slightly. Leaving the teas near each other amounted to a sort of cosmic misdemeanor. Wheeler quipped that he might hide any evidence of it if, before the teas began to equalize temperatures, he dropped them into a black hole. Inside the event horizon, the teas and the associated entropy would be forever cut off from the rest of the universe and thus would be unable to contribute to its entropy.

The student to whom he had made this small joke was Jacob Bekenstein. Bekenstein thought about it and realized that he was bothered by the presumption that black holes were somehow exempt from the second law of thermodynamics. Some weeks later he told Wheeler that he was wrong—that his small crime would, in fact, be detectable even if he dropped the two cups of tea into a black hole. The entropy, Bekenstein said, would not disappear from the universe at all. Rather, it would increase inside the black hole's event horizon, and this entropy would be represented to the outside universe in an increased surface area of that horizon.[8]

Stephen Hawking had initially thought Bekenstein's hypothesis absurd. In 1974, though, Hawking confirmed it and did quite a bit more. By formulating a "partial marriage" of general relativity and quantum mechanics, he showed that black holes radiate, and they continue to radiate even when their spinning has slowed and after it has ceased completely. Hawking suggested that, in this manner, a black hole would evaporate, and when enough of its mass was lost, what remained would explode violently.[9]

What has come to be called *Hawking radiation* would be driven by vacuum fluctuations. Again, the wave–particle duality allows us our choice of representations, and Hawking radiation may be easier to understand if we imagine it as particles. Imagine a vacuum fluctuation creating two virtual photons just above a black hole's event horizon—one having negative energy, the other having positive energy. Normally they would re-join and annihilate each other almost instantly. But here, tidal gravity pulls them farther apart. The negative-energy photon falls below the horizon, and the positive-energy photon escapes the hole altogether, carrying with it some of the black hole's energy (and, as Einstein's famous equation demands, some of its mass). The escaped photon, having added energy, is no longer virtual: it is real.

The same process would operate for all types of radiation, including gravitational radiation. To be sure, this particle-by-particle evaporation would be a slow process. A black hole the size of the Sun would take far longer than the age of the universe to evaporate, and for that reason its evaporation would mean little to astrophysicists and astronomers working to understand the life cycles of stars. Nevertheless, Hawking radiation supplied a missing piece to a rather large puzzle. Because negative energy falling into a black hole counters radiation flowing out of it, Hawking radiation is the mechanism by which black holes reach thermal equilibrium with their environment and thus abide by the laws of thermodynamics as we know them. Negative energy, although a very small presence, is necessary to the explanation of how black holes fit into our universe.

In 1978, physicist Lawrence Ford discovered that nature was extremely stingy with negative energy. If we produce more negative energy, we cannot produce it for as long a period. If we produce it for a longer period, we cannot produce as much. In any case we cannot produce much to begin with. These severe restrictions were called "quantum inequalities," and during the 1990s they were sufficiently proved and refined that they could be brought to bear on other problems—like the maintenance of wormholes. In 1996, Ford and Thomas Roman found that quantum inequalities would allow enough negative energy to sustain small wormholes indefinitely, but these would be very small indeed, with throat radii only slightly larger than a Planck length, the smallest length that has meaning.

Although larger wormholes—that is, wormholes that might be traversable by humans—were not quite forbidden by quantum inequalities, they would present a fairly daunting engineering challenge. In his 1995 textbook, Visser calculated that "building a one metre wide thin-throat traversable wormhole requires about one Jupiter mass of exotic negative energy matter."[10] Note that the acquisition of this amount of negative energy would be far more challenging than the acquisition of the same amount of positive matter required to construct the mass shell in Chapter 2. A Jupiter mass of negative energy would have a magnitude equivalent to the total energy generated by ten billion stars in a year. According to the quantum inequalities, such an undertaking would also take considerable finesse. The energy would need to be confined to a "skin" enclosing the mouth with a thickness one-millionth the width of a proton.

In 1999, Ford and Roman found that quantum inequalities imposed restraints that were still more severe. We might "borrow" negative energy, but we would be expected to pay it back with interest. If we use a pulse of negative energy, a pulse of positive will follow; and the longer we can make the delay between the pulses, the greater the "loan" time will be, and the stronger the positive pulse will be.[11] When Ford and Roman imagined schemes to separate the pulses (and so attempt to isolate the negative energy), they found that all were, for various reasons, unworkable; and in many cases the very act of separating the pulses produced positive energy. It was clear that a little negative energy is necessary to making sense of the universe; but it was beginning to seem that a lot—enough to hold open the mouth of a traversable wormhole—was impossible.

CHAPTER NINE

A PARADE OF TIME MACHINES

IN WHICH PHYSICISTS IMAGINE A VARIETY
OF TIME MACHINES

S hortly after the Aspen workshop, Matt Visser attempted to take a larger view of the question of time machines. His conclusions appeared in a 1993 paper that was the closest anyone had come to defining and characterizing positions as they had developed thus far.

Taking Stock

It was clear that opinions fell roughly into four camps. Members of the first appreciated the participation of respected physicists and the tacit approval of the editors of *Physics Review D*, but they thought the subject still smacked of science fiction. There was no empirical evidence for traversable wormholes, let alone traversable wormholes made into time machines. Pending further discoveries, then, it seemed reasonable to put the matter aside. But as far as Visser was concerned, they were tossing out the baby with the bathwater. In dismissing time machines they were also dismissing a large set of interesting questions. Visser felt that the upshot of their position was to prohibit work in some very exciting areas. He called it the "boring physics" conjecture.

The second position was the one taken most famously by Novikov and adopted provisionally by several others. Novikov held that both traversable wormholes and traversable closed timelike curves were possible—or at least that no one had proven them impossible. Moreover, he said, there was reason to believe that natural laws would force closed timelike curves to be self-consistent. Pastward time travel need not violate causality.

Visser agreed with the first part of Novikov's argument but found the second untenable. Regardless of the proximate cause of any case of self-consistency—in the example of David Lewis's "Tim" (whom we encountered in Chapter 5), that cause might be a braced shoulder slipping or a particle of sand in a rifle barrel—Novikov's supposition implied the involvement of many factors working to preserve a specific course of events, all somehow coordinated and put into motion by a mysterious and ill-defined force.[1] The sum over histories struck Visser as a rather weak mechanism and an unlikely candidate for that force. He suggested that it was ad hoc, and in the rough-and-tumble prose style rather characteristically his own, he called it the "Novikov consistency *conspiracy*."

There was, for Visser, another reason to think that self-consistency was naïve. Ideas that made consistency a necessity were nothing more than ways of requiring that the universe fit human predisposition. They were, he said, "firmly wedded to the notion of spacetime as a four-dimensional Hausdorff differentiable manifold"[2]—roughly speaking, a four-dimensional space in which points can be separated by "neighborhoods" and on which one can perform calculus.[3]

By the 1990s, many physicists had come to regard Minkowski's spacetime as chemists regarded Bohr's "Solar System" model of atomic structure—that is, as a useful way to begin to think about the subject, but fundamentally inaccurate and potentially misleading. It was possible that such spacetime was merely "local." Some thought that our universe represents a single branch of a much larger, much older entity, a *multiverse*, that generates new universes continually. Accordingly, Visser said, "One could question the naïve notion that the 'present' has a unique fixed 'past history.' After all, merely by adding a time-reversed 'branching event' to our non-Hausdorff spacetime one obtains a 'merging event' where two universes merge into one. Not only is predicibility [*sic*] more than somewhat dubious in such a universe, but one appears to have lost retrodictability as well."[4]

Unmaking the Past

The idea of a spacetime that branches not only into the future but also into the past is a radical concept, but in outline at least, it is not a particularly new one. In the eleventh century the Roman Catholic Church was riven by debate, and theologians who had received training in logic, the third part of the group of studies known as the trivium (the others were grammar and rhetoric), applied their skills to realms long thought beyond the reach of human reason. Berengar of Tours worked to solve the mystery of the Eucharist; Roscelin of Compiègne, that of the Trinity. They and thinkers like them came to be called dialecticians, and they brought about a historical moment during which there was reason in faith as well as faith in reason—a historical moment, some would say, quite unlike our own.

As might be expected, more traditional thinkers regarded such inquiries as prideful and profane. Among them was an Italian monk and author named Pietro Damiani—or, as his English translators called him, Peter Damian. Peter held great disdain for the dialecticians and did not bother to hide it. "These men," he wrote, "lose their grasp of the fundamentals of simple faith as a result of the obscenity produced by their dull tricks; and, still ignorant of those things boys study in school, they heap the abuse of their contentious spirit on the mysteries of God."[5]

Nonetheless, when pressed, Peter could summon formidable

rhetorical skills, many of which are demonstrated in a letter called *De omnipotentia dei*. It begins by arguing that God is capable of undoing events—in Peter's example, the loss of virginity—and goes on to suggest that God may change the past such that—again in Peter's example—Rome was never made. It is worth noting that arguments that God is *unable* to change the past had been made by Aristotle, Saint Jerome, and Thomas Aquinas, all of whom suggested that the prospect introduced an inconsistency. To this, Peter countered that consistency may be thought of as a sort of natural law. Because God created natural law, he is not bound by it, and in fact his miracles are miracles precisely *because* they defy natural law. If God is not constrained by natural law, it follows that he cannot be constrained by consistency.

However, Peter's whole argument is subtler than this, depending for its details upon the proposition that God is not merely eternal, but present, for lack of a better word, *simultaneously* in all moments past, present, and future.[6] Indeed, Peter's God seems to pervade and imbue something very like a block universe: "He contains within himself the march of all time; and just as he holds within himself all times without their changing, so too within himself he encloses all spaces despite their spatial difference."[7]

Peter nowhere asserts that God actually *has* changed the past—and, in fact, argues that for him to do so would be acting against his nature, which is generative, not destructive. He argues only that God has the power to do so. Neither does Peter speculate upon the effects of such an action. Anyone else, though, might find such musings difficult to resist. We are tempted to ask, if God unmade Rome, would its histories vanish as well? Its ruins? It would be as though a force outside the block universe removed one section of spacetime and replaced it with another. Such ideas lead beyond the tenseless theory of time and beyond the block universe, to conceptions of reality far stranger—conceptions in which, as Visser would later say, "all hell breaks loose."[8]

A theoretical spacetime that branches into the future as well as the past had been explored, among other places, in a 1979 paper by Roger Penrose. To appreciate it, we must introduce what Eddington called the *arrow of time* and its associated problems.

Imagine a film in which a pot of geraniums is nudged from a balcony ledge, drops through the air and falls sideways, strikes the pavement below, and explodes in a spray of pottery shards, soil, and geraniums. Nothing about the film seems particularly remarkable. But play the same film in reverse. The soil collects itself, pulls together the suddenly uprighted geraniums, and the pottery shards draw each other together in the shape of a pot, sealing the soil inside it. This suddenly compact assembly rockets upward, slows and straightens itself, and finally comes to rest on the balcony ledge. Now the film seems strange, and it seems so because it violates our notions that time ought to behave in accordance with an "arrow" that points in one direction.[9]

Time's Arrow

Although it is true that the event described in the previous section is extremely unlikely, it is also true that no local physical law forbids it. In fact, the laws associated with three of the four fundamental natural forces—the gravitational force, the strong nuclear force, and the electromagnetic force—are utterly indifferent to time's arrow. Likewise, *quantum mechanics*, the set of laws that govern the universe on very small scales (with a single exception to be discussed shortly), is believed to have no preferred direction in time.[10] The question becomes, why should the universe on large, macroscopic scales—the scales of humans and flowerpots—show a preferred direction for time, and the same universe on small scales—the scales of atomic or subatomic particles—show no preferred direction? Astronomer Edward Harrison likens the situation to an oil painting. Examining a small section of canvas reveals only daubs of paint, with no preferred "up" or "down." To gain a sense of the painting's proper orientation, to know which way to hang it, you must stand back far enough that your vision takes in a larger area.

One of the more radical approaches to understanding quantum mechanics is the *Everett interpretation*, named for Hugh Everett (who, as it happens, was another of Wheeler's students, and about whom we will learn more in Chapter 10). Briefly, the Everett interpretation claims that the universe branches into two or more universes whenever a measurement takes place, and it branches as many times as there are possible outcomes of the measurement. Now we are prepared to let all hell break loose. If quantum mechanics has no preferred direction in time, and if Everett is right, then it follows that there should be *as much branching into the past as into the future*.

This rather startling idea, in rough form, yields a picture of a "block multiverse" whose worldlines represent whole universes and branch in all directions, downward as well as upward. What this would mean on the scale of human experience is difficult to visualize, and it may be that we would be unable to distinguish whether we were in a forward-tending or backward-tending branch. At any rate, in the 1979 piece, Penrose conjured this bizarre image only to claim that it was rooted in an error. The assertion that the quantum realm recognizes no direction in time is not quite true. Particle physicists have discovered one subatomic process that *does* acknowledge time's arrow.

Most subatomic particles are unstable, and after a brief period they decay into smaller particles. A part of the decay of a particle called the K^0 minus meson is in accord with time's arrow. The effect is so minute that it cannot be measured and must be inferred, and so far as anyone has been able to tell, it plays no significant role in any of the processes that govern the behavior of matter. Still, it has been confirmed many times over. It is real. What is curious about the effect is that it is so small that it seems hidden. It is as though a single daub of paint on Harrison's oil painting, examined under a microscope, were revealed to contain, in miniature, an image of the entire painting.

Penrose had no explanation for this effect, but he observed that something similar occurs in biology. Most of Earth's animal life shows bilateral symmetry: the body of an animal can be cut lengthwise into two halves that are reverse images of each other. Curiously, this life is governed by a structure that has no corresponding symmetry—a helix that is decidedly right-handed. As to the time-asymmetric aspect of the decay of the K^0 minus meson, Penrose admitted that he did not know what, if anything, it might mean. He did say, rather provocatively, that it seems a case of nature "trying to tell us something."[11]

One of the most profound questions of physics is what the connections among these several arrows of time are. The K^0 minus meson seems part of an answer; at the very least it suggests that time is asymmetric on all scales.

The third position on time travel, however, disregarded this suggestion and assumed the viability of the Everett interpretation of quantum mechanics, along with its implication of a great—perhaps infinite—number of universes branching forward and backward in time. Whatever else it was, the view was not boring. But it required physicists to assume that fundamental principles of cause and effect were useless, and again raised the specter of causality violation. As Visser noted, it trumped Hawking's nightmare of a universe unsafe for historians with a universe utterly *unintelligible* to them, because it undermined the very idea of a unique history. In an earlier paper, he observed, "This possibility is exceedingly unpleasant, and rather worse for the state of physics than any of the (relatively conservative) ideas."[12] For all these reasons he termed the position the "radical rewrite conjecture."

For Visser, though, there was a middle road, and it had been blazed by Hawking. The chronology protection conjecture allowed for the existence of wormholes, even traversable wormholes. The protection mechanism engaged, Visser pointed out, not when the wormhole was constructed, and not when one mouth was accelerated to relativistic velocities. It engaged when the traveling mouth was being returned to the vicinity of the stationary mouth. Visser expected that, at the moment the wormhole became a time machine—or rather, again owing to quantum indeterminacy, a brief interval *before* that moment—a number of physical mechanisms came into play: "vacuum polarization effects," "wormhole disruption effects," and a "gravitational back reaction." Any one of these was sufficient to destroy the would-be time machine. Taken together, they comprised a whole set of fail-safe mechanisms with which the universe protected itself against causality violations. He called it a "defense in depth."[13]

In the months and years after the Aspen workshop, several physicists began to imagine ways in which such mechanisms might be mitigated.[14] One was to twist spacetime. In part because no one had a practical means to measure the topology of actual space, most cosmologists considered the universe curved, open, or flat, and little thought had been given to alternatives. But in the 1990s, in some measure because of the influence of mathematicians, cosmologists began to think about twisting spacetime into a myriad of shapes. And a few began to imagine that some of those shapes might allow closed timelike curves.

Time Machines in Other Spacetimes

If a physicist wishes to introduce a new approach to a problem but is unprepared to put forth a full-fledged model, he might propose a "toy model"—that is, a model that omits certain features of reality but includes others considered important and relevant. In the

1990s a number of physicists imagined toy models of spacetimes in which time travel might be possible.

In November 1992, David Boulware at the University of Washington in Seattle found that, for a sufficiently massive field in the spacetime created by Gott's cosmic strings, a closed timelike curve would survive a potentially destructive back reaction caused by the buildup of vacuum fluctuations. In 1988 William Hiscock had suggested that Thorne consider the effect of vacuum fluctuations. In 1995, he and Tsunefumi Tanaka, both at Montana State University, examined a field like that described by Boulware, and they also concluded that the back reaction might be minimized. In 1996, Sergei Krasnikov at the Central Astronomical Observatory at Pulkovo, St. Petersburg, proposed a toy model of a two-dimensional spacetime in which closed timelike curves were stable. And in 1997, Sergey Sushkov at Kazan State Pedagogical University showed that they might also be stable in Misner space, a hypothetical spacetime in which closed timelike curves do not exist originally, but eventually develop.[15]

A Two-Wormhole Time Machine

Working independently, Matt Visser and Maxim Lyutikov suggested in 1994 that there might be a way in which the back reaction could be made smaller using *two* wormholes.[16] Suppose we begin with a single wormhole. It has one mouth on Earth and another mouth in deep space. The mouth on Earth is time-dilated one-half hour with respect to the one in deep space. If we send a radio signal through the mouth on Earth, it emerges from the mouth in deep space one-half hour in the past as viewed from Earth. Now suppose that via normal space, the deep-space mouth is one light-hour from Earth. This means that a signal sent round-trip—that is, outward bound from Earth through the wormhole and inward bound through normal space—would, when the adding and subtracting are over, return one-half hour after it left. We would have a vivid demonstration of special relativity but, since the signal would arrive too late to meet its earlier self, no time machine.

Now introduce a second wormhole, laid head to toe alongside the first—that is, with the mouth in deep space time-dilated one-half hour with respect to the mouth on Earth. To avoid complications, assume the distance in normal space between the two deep-space mouths to be so small that we may disregard it. A signal sent round-trip from Earth—outward bound through the first wormhole, across the brief span of normal space, and inward bound through the second wormhole—would arrive at the same moment it left. Moreover, by further time-dilating the mouths, we might arrange for the signal to return *before* it left. In either case we have created a time machine. What is significant about this arrangement is that the wormholes are a time machine only when used together. The advantage of paired wormholes over a single uninterrupted wormhole time machine is that the vacuum fluctuations are defocused when they emerge from the time-dilated mouths, thus taking pressure off the whole system.

The Alcubierre Bubble

The closed timelike curves in the cases described thus far in this chapter, once created, would in principle be stationary in the space-time in which they lived. In 1994 a young physicist named Miguel Alcubierre suggested a rather more dynamic model.

Alcubierre was born in Mexico City; in 1990 he moved to Great Britain to begin graduate work at the University of Wales at Cardiff. It was there that he produced a paper describing a hypothetical spacecraft embedded in a bubble of warped spacetime.[17] The bubble might be given any trajectory desired, and space along that trajectory would be infinitely elastic; that is, space ahead of the bubble could contract by any amount, and space behind it could expand by any amount.

Alcubierre acknowledged that such a project would necessarily require large negative energies, but for the purposes of his paper that did not matter. His intent was merely to show that the necessary stretching and squeezing was allowed by general relativity. Because the bubble itself would be no physical entity, but rather a region of spacetime, it would be able to move at supraluminary speeds. The spacecraft would be constrained to subluminary speeds relative to the spacetime within the bubble, but it would be carried *by* the bubble, and so (with the bubble) it could move at supraluminary speeds relative to the spacetime outside it.

Alcubierre concluded the piece with two observations. First, it seemed obvious that such propulsion merited the name "warp drive" from science fiction. Second, and perhaps more intriguingly, it implied the possibility of pastward time travel.

A year later, Allen Everett at Tufts University noticed Alcubierre's paper. He had already given thought to a sort of supraluminary travel when, some years earlier, he had written on tachyons, the hypothetical (and, in the view of most, nonexistent) particles that travel faster than light and cannot travel slower than light. Modifying Alcubierre's model slightly, Everett showed how it might be made into a time machine.[18]

Imagine the spacecraft's worldline in the center of two superimposed light cones. One light cone encloses spacetime within the bubble; the other encloses spacetime outside it. Tip the light cone representing spacetime within the bubble clockwise until the spacecraft's worldline passes through the surface of the "outer" light cone, and keep tipping it clockwise until it passes the 135-degree mark and so passes again through the surface of the "outer" light cone, this time going in the other direction. Remove the "inner" light cone—in effect, burst the warp bubble—and the spacecraft is in its own past.

Everett noted that the model bore some resemblance to Gott's cosmic-string time machine, although it was perhaps more realistic. Whereas the cosmic strings extended to infinity, the warp bubble was "compact"; it could be constructed in a finite region of spacetime. Everett acknowledged that the stretching and shrinking would need exotic matter—the same stuff needed to hold open wormhole mouths—and he noted that the quantities necessary were, to put it mildly, unrealistic. A bubble two hundred meters in diameter would require a negative energy with a magnitude equivalent to ten billion times the mass of the observable universe.[19] He also observed that there would be a chronology horizon and an associated back reaction, which, if Hawking was right, would destroy this closed timelike curve as quickly and effectively as it would destroy the others.

The Krasnikov Tube

About the same time, Krasnikov discovered what was perhaps a more fundamental flaw in the scheme.[20] When the bubble is moving at supraluminary speed, its interior is isolated from its own surface and from the universe outside it. Nothing—no matter, no energy, no signal—can pass through that surface in either direction. This means that the spacecraft cannot control the bubble from inside it, and it was difficult to imagine how the bubble might be controlled from outside. Any control would need to maintain some sort of causal connection, and once the bubble was traveling faster than light with respect to the universe outside it, nothing in that universe could keep pace. So, in the same paper, Krasnikov proposed a modification.

The spacecraft travels outbound at near light speed. Although special relativity limits the effect that the spacecraft can have on the space ahead of itself, the spacecraft can affect *all* the space through which it has already passed. Krasnikov suggested that, if the spacecraft could by some means shrink that space, it might make its inbound trip through it. Furthermore, if that space were shrunk sufficiently, the spacecraft could arrive mere moments after it left. It could not, however, arrive *before* it left, and for this reason Krasnikov proclaimed his model no threat to causality.

On this last point others were not so certain. In 1997, Everett and Roman realized that the shrunken space through which the spacecraft traveled on its inbound leg—what they dubbed a *Krasnikov tube*—was, in purely functional terms, a traversable wormhole. They concluded that for this reason two tubes—recalling Visser and Lyutikov's scheme of twinned wormholes—might indeed violate causality, although they also noted that, as with Alcubierre warp drive, maintaining them would require unrealistic amounts of negative energy.[21]

Meanwhile, exactly what happened on the chronology horizon was a matter of continuing controversy. All concerned were aware that a theory of quantum gravity would be necessary to tell the whole story. In its absence, physicists made do with approximations: specifically, semiclassical physics and semiclassical quantum gravity. *Semiclassical physics* is the set of laws that regard spacetime according to Newton and Einstein (that is, well defined and classical), but treat matter according to quantum mechanical laws (that is, subject to uncertainty and quantum fluctuations). *Semiclassical quantum gravity* treats gravitational fields as well defined and classical, and matter fields as quantum.

Among the work that called upon these approaches was research reported in a 1993 paper by Li-Xin Li, Jian-Mei Xu, and Liao Liu—all at Beijing Normal University. It voiced a small dissent from Hawking's description of the horizon, suggesting that spacetime there might take the form of a "quantum barrier" that would be stable against the back reaction.[22]

The first author of the paper—Li-Xin Li—had, as an undergraduate, gained a reputation for marathon sessions of Go, or *wei ch'i*, the two-player board game that, in ancient China, was regarded as one of the four skills—along with brush painting, poetry, and music—to be mastered by a cultured man, in part because it rewarded patience and balance over aggression. Li had patience—and persistence—to spare. He also had audacity. He was still a master's candidate when he struck out on his own, suggesting in a single-author paper that a "spherical reflecting mirror" positioned between the mouths of a wormhole time machine would deflect the vacuum fluctuations so as to prevent the infinite buildup of energy and preserve the closed timelike curve.[23]

By 1996, Li was in the graduate program at Princeton, where he proposed a similar scheme, arguing that, if an "absorbing material with appropriate density" could be placed between the wormhole mouths, it might smooth out those same fluctuations and likewise preserve the closed timelike curve.[24] In time, Li would become the most tireless opponent of Hawking's chronology protection conjecture. Between 1994 and 1998 he published four papers in *Physical Review D* and *Physical Review Letters*, each suggesting that instabilities posited by Hawking and others were exaggerated or could, by ingenious means, be circumvented.

Novikov Finds Precedent for the Self-Consistency Principle

All these models concerned what we might think of as engineering particulars. As they were put forth and debated, one physicist was still seeking a more expansive view. It was, of course, Igor Novikov.

Since Aspen, Novikov had sought to ground self-consistency in a more fundamental tenet, and by 1995 he had found one: a sort of classical analogue to the sum-over-histories interpretation called the "principle of least action." Its first formulation, in 1746 by French scientist Pierre-Louis de Maupertuis, asserted that all natural processes are economical. In its rough form, such a claim is not particularly surprising. We all know, for instance, that water running downhill will take the steepest path. But when we examine such processes more closely, we may find surprises.

Richard Feynman told of his own introduction to the principle of least action. His high school physics teacher had said, "You look bored; I want to tell you something interesting."[25] He asked Feynman to imagine that someone in a gravitational field threw something. He called it a "particle"; we may think of it as a baseball. It traveled along a parabolic arc until, at a given point, it hit the ground. If we allow the time the baseball takes to travel to vary, there are an infinite number of paths that it may take to that point: a straight line, a higher arc, a slightly more shallow arc, and so on. If we lock in a specific time for travel, the ball may still take an infinite number of paths, although it will not travel freely: we will have to direct the ball's motion and adjust its speed. But for all these paths—directed and undirected—we can calculate the ball's kinetic energy and its potential energy. What Feynman's teacher showed him, and what Feynman found fascinating, was that, for the actual free-traveling path, the difference between those energies was smaller than that for any other path.

Every branch of natural science has "minimal" principles, and all are likely to give us pause. One of the first was formulated by Pierre de Fermat about 1650. Called "Fermat's principle of least time," it asserts that, of all possible paths between one point and another, light always takes the fastest.

The path that light takes through a single medium will be, as we might expect, straight. But because light travels more quickly through some mediums than through others, a path through more than one medium may be otherwise. For instance, light travels more quickly through air than it does through glass. Suppose that, in the air on either side of a pane of glass are two points, and the shortest path between them is a diagonal through the glass. Strange as it may seem, light traveling between the points will not follow that diagonal. Instead, it will travel through the air at an angle nearer the surface of the glass than the diagonal, and it will strike the glass at a point farther away than the place that the diagonal struck. There it will bend and travel through the glass at an angle nearer the vertical than the diagonal. At journey's end, the light will have spent more time in the medium in which it travels fastest and less time in the medium in which it travels slowest, and its total travel time will be less than it would have been along the diagonal path or, indeed, any other.

It is difficult to think about these phenomena without endowing them with will and foresight. Even in the macroscopic, classical realm—the world of waterfalls and baseballs—we are tempted to ask the question that we asked of the electron in the double-slit experiment: how does it "know"? Novikov did not claim to have an answer, but in 1995 he and four colleagues brought the principle of least action to bear on closed timelike curves.[26] In a thought experiment involving a "hard sphere" (in essence, the billiard ball of earlier papers), they found that even in dealings with time machines, nature was economical. Of all the possible paths along the closed timelike curve, those that were self-consistent were also the most efficient.

As strange as the principle of least action may seem, it was formulated 250 years ago and features in our everyday existence. In applying it to closed timelike curves, Novikov and his colleagues had made time machines seem a bit more—for lack of a better word—"natural."

The Roman Ring

The physicists thinking about closed timelike curves in the 1990s were dispassionate in approach, but not always in presentation of their work.[27] Some regarded closed timelike curves as a sort of engineering challenge, and the putative mechanisms of chronology protection as obstacles to be overcome. Among these were Li, Novikov, and Israeli physicist Amos Ori. But they were a minority. Most physicists would have preferred a universe in which time

machines did not exist. Among them—and perhaps the most prolific of them—was Matt Visser. In fact, it was precisely *because* Visser
was uneasy with the idea of time machines that it is of some note
that, after the Aspen workshop, like a man who cannot resist
scratching a sore, he worked to develop several original models that
violated—or threatened to violate—causality.

One of these, described in Visser's 1995 textbook, was a "longi-
tudinally spinning cosmic string"—a sort of thinner, distant cousin
to the Van Stockum cylinder and Tipler cylinder, and arguably more
realistic than either. Another, described in a 1997 *Physical Review D*
piece, was a carefully arranged set of traversable wormholes.[28]
Visser saw no reason to stop at one or even two. In fact, he noted
that the more wormholes one uses, the smaller the back reaction
becomes and the more stable the whole system becomes. In
essence, the idea was not new. The scheme that Thomas Roman
had suggested to Morris in 1986 had used two wormholes. Because
Visser's model was in some ways its embellishment, it was only
appropriate, in his view, that he call it a *Roman ring*.

Hawking Gives Some Ground, and Takes Some Back

What, then, of Hawking? By 1998 he had seen five years' accumulation of counterexamples to his assertion that the back reaction
was the mechanism by which chronology protection was enforced.
In February of that year he gave ground—but not much. He sur-
rendered the mechanism but held fast to the conjecture itself. With
student Mike Cassidy, Hawking brought Feynman's sum-over-
histories interpretation of quantum mechanics to bear on the problem and developed another protection mechanism—one that he
considered more durable.[29]

Recall that Feynman's interpretation acknowledged all histories
of a given particle. Cassidy and Hawking observed that these
included even those histories that travel faster than light or back-

ward in time. Using a simple model of spacetime that was mathematically equivalent to models that allow closed timelike curves, they calculated the probability that a closed timelike curve would actually form, and found it to be infinitesimally small. Although Li published a piece that challenged the idea a year later,[30] in the journals at least, Cassidy and Hawking's prediction was met mostly with silence. The reason may have been that sum over histories was rooted in probabilities, and so as a protection mechanism it was less than ideal. Those desiring chronology to be safeguarded would have preferred an inviolable law.

The Limits of Knowledge

By the end of the 1990s, most physicists working in the area agreed that a fully satisfying chronology protection mechanism had not been found. What did this mean to the larger question—that is, the viability of time machines? The answer was likely to be shaded rather differently, depending on who was doing the answering. Li and a few others asserted that it meant that time travel had yet to be proved impossible. This was cautiously negative phrasing, firmly in the Sagan mode, and barely disguising an unvoiced wish that time travel might one day be realized. Visser, Hawking, and most of the others—a clear majority—said that it meant only that they lacked the necessary tools. Neither semiclassical physics nor semiclassical quantum gravity was sufficient to identify a mechanism that protected chronology.

There was one point, however, that all could agree on: the chronology protection mechanism, if it existed, lay in the chronology horizon. As it happened, that location was precisely the problem. In Newtonian mechanics, the source of gravity is mass; in general relativity, the source of the gravitational field is the *stress-energy tensor*. In 1997, three physicists—Bernard Kay, Marek Radzikowski, and Robert Wald—demonstrated that one could not define a stress-energy tensor on the chronology horizon and that, at places on the horizon, the semiclassical Einstein equations simply do not hold.[31] What had long been suspected was proved. There was no way to understand what was going on at the horizon other than through a full theory of quantum gravity.[32]

This was a humbling, and in some ways an unwelcome, bit of news. The idea that cause precedes effect is commonplace. It is also so fundamental to Newtonian mechanics, special relativity, and quantum field theory that, in those subject areas, any hypothesis purporting that effect might precede cause would be rejected out of hand. Now it seemed that causality itself, long taken for granted, had yet to be grounded in physical law.

CHAPTER TEN

TIME TRAVEL IN ANOTHER KIND OF SPACETIME

IN WHICH IT IS PROPOSED THAT, IF OUR UNIVERSE IS BUT
A SINGLE BRANCH OF MANY, PASTWARD TRAVEL
NEED ENGENDER NO PARADOX

This web of time—the strands of which approach one another, bifurcate,
intersect or ignore each other through the centuries—embraces every
possibility. We do not exist in most of them. In some you exist
and not I, while in others I do, and you do not,
and in yet others both of us exist.
— Borges, "The Garden of Forking Paths"[1]

David Deutsch is a visiting professor of physics at Oxford. He is
recognized internationally, among other reasons, for his groundbreaking work in the field called quantum computation—that is,
computers that exploit the effects of quantum mechanical phenomena. He is in his early fifties but might easily be mistaken for
someone much younger. He has a gentle manner, yet, on subjects
about which he feels strongly—and there are many—he becomes
decidedly impassioned.

Deutsch's formal training was like that of many physicists we
have met thus far. His early work concerned quantum field theory
in curved spacetime; he studied under Dennis Sciama at Oxford and
did postdoctoral work with Wheeler. Like many in this book,

Deutsch publishes in *Physical Review D*. What makes Deutsch different is that he also writes for journals in philosophy and symbolic logic. He is interested in the philosophy of science—meaning the study of the foundations and implications of science, and exactly how scientific ideas are shaped—and he is not shy about bringing such concerns into his work in physics. A glance through the footnotes of his papers reveals not only the usual references to Wheeler and Hawking, but also citations to philosophers Karl Popper and Jacob Bronowski.

In the spring of 1991, Deutsch was working at the Oxford University Mathematical Institute. It had been a year since the physicists whom Thorne called the consortium had published the paper that summarized work on closed timelike curves and concluded, among other things, that self-consistent past histories meshed with the sum-over-histories formulation of quantum mechanics. Deutsch followed this early give-and-take with considerable interest, and a measure of chagrin. As far as he was concerned, they were all wrong, and they were all wrong because they were thinking about the problem in the wrong way to begin with. His point was that, although all parties agreed that whatever was happening at the chronology horizon was a quantum mechanical process, they persisted in thinking about it in classical terms.

The Smallest Time Machines

Most physicists would agree that the nature of physical reality is fundamentally quantum. For this reason, closed timelike curves may be quite common; if quantum foam is real, there may well be submicroscopic time machines everywhere. To be sure, they are very small (10^{-35} meter across) and very short-lived (they stay open for 10^{-43} second—that is, about one Planck-Wheeler time), so they would loop pastward only that far. Nonetheless, it was obvious to Deutsch that the questions surrounding closed timelike curves demanded a quantum mechanical approach. But exactly what a "quantum mechanical approach" might be was an open question in 1991 and remains one today.

Recall that knot of quantum weirdness—that measuring one part of a quantum system alerts other parts of the system even when the parts are too far apart for a signal to travel the distance between them. An experimenter sees an electron passing through the left-hand slit, and the electron seems somehow to coordinate its behavior with another, invisible electron passing through the right-hand slit. The Copenhagen interpretation explains the result by claiming that the wave function or state "collapses" from wavelike behavior to particle-like behavior at the moment it is measured.

Despite the challenges that it presents to common sense, the Copenhagen interpretation is the favored interpretation of quantum mechanics today, and it was most favored in the 1950s, when a young man named Hugh Everett was doing graduate work at Princeton. Everett was working under Wheeler, who recalled him as "an independent, intense, and driven young man."[2] Everett considered the Copenhagen interpretation ad hoc, and in that sense he was in the company of many—but he broke ranks when he offered an alternative. In the spring of 1957 he submitted a thesis describing what he termed a "relative state" formulation of quantum mechanics—or, as it became known, the "many universes" interpretation.

Everett's Many Universes

Everett's interpretation dismissed what many regarded as the metaphysical baggage of the Copenhagen interpretation. It had no need for collapses into reality. The reality, in Everett's view, was already there. Everett claimed that the act of shooting an electron through the first slit literally splits that electron into two electrons, and the universe along with it into two universes. There actually is another electron passing through the right-hand slit, and it is every bit as real as the one we detected passing through the left-hand slit. It is invisible to us, quite simply, because it exists in that other universe. In that universe an experimenter detects an electron passing through a right-hand slit, which seems somehow to coordinate its behavior with an invisible electron passing through a left-hand slit—the electron we detect in our universe. What the double-slit experiment measured, then, was an "interference effect" by which these two universes impinge upon each other.

Wheeler gave the work his stamp of approval, and when Everett's paper was published in *Reviews of Modern Physics*, Wheeler composed a short piece to accompany it. He once joked that he set aside one Tuesday a month to disbelieve the idea,[3] but in truth his reservations were deeper. Wheeler intended his endorsement only to signal that Everett's idea was important—not necessarily correct—and he was mildly unsettled when, for many years, it was called the Everett–Wheeler interpretation.

One of the paper's first readers was physicist Bryce DeWitt at the University of Texas at Austin. DeWitt was skeptical, but intrigued. Everett and DeWitt began an exchange of letters, and Everett answered DeWitt's objections point by point. When DeWitt, in a last-ditch appeal to common sense, said, "I just can't feel myself split," Everett replied that those who objected to Galileo's ideas of an Earth in orbit around the Sun said that they could not feel Earth move beneath their feet.

DeWitt was converted, and in due course he would put so much work into developing the idea that his surname came to replace Wheeler's after the hyphen. It is DeWitt who coined the term "many universes" as applied to the interpretation, and DeWitt who has described, unapologetically, its mind-boggling implications: "Every quantum transition taking place on every star, in every galaxy, in every remote corner of the universe is splitting our local world into myriads of copies of itself."[4]

The paper that Everett had published in 1957 gave, as its return address, "Weapons Systems Evaluation Group, the Pentagon." When he submitted it, he had already left Princeton and was working on various classified projects involving, among other things, early computers. Meanwhile, the interpretation itself was mostly ignored, and the few who thought of it at all considered it heretical. Physicist Philip Pearle, in a marvelous bit of understatement, termed it "uneconomical."[5] But the competing interpretations presented their own problems, and the result was a situation that allowed astrophysicist Steven Weinberg to call the Everett interpretation "a miserable idea except for all the other ideas."[6] Then, in the 1960s and 1970s, when some believed that quantum mechanics might rescue theoretical physics from the spacetime singularities implied by general relativity, quantum mechanics as applied to the whole universe became of interest. It was in this application, a field called quantum cosmology, that the Copenhagen interpretation ran into trouble.

The Problem of the Observer

The Copenhagen interpretation asserts that an isolated quantum system, such as the electron traveling through the double-slit apparatus, is a collection of superimposed quantum states—or, in the preferred phrase, a "wave function." As long as the wave function is unobserved, it may be imagined as a complex and changing pattern of waves, like ripples on the surface of a lake. But at the moment it is measured, all the waves collapse into a single wave. According to the Copenhagen interpretation, the act of measurement actually causes the collapse. Of course, the act of measurement requires a measuring apparatus. If classic-sized physicists are to use it, the apparatus must be much larger than the effect it measures. Nonetheless, because it is composed of subatomic particles the apparatus is subject to laws of quantum mechanics. As the quantum system reacts with the apparatus, so the apparatus reacts with the quantum

system. And because it is impossible to say exactly where the apparatus ends and the effect it measures begins, the apparatus and the
effect may be considered part of a larger quantum system. But until
it is measured, this system, too, must be a collection of superimposed
states. So the measuring apparatus can be said to have collapsed the
wave function only when it is measured by a *second* apparatus, and
so on, to an infinite regress of measurements.

Thus arises what is called the "problem of the observer." In a
finite system, we can safely ignore it. Whatever system we decide to
measure, there will always be someone or something outside it that
can collapse the wave function. But suppose, as in quantum cosmology, we wish to treat the universe itself as a quantum system. No
observer or measuring apparatus can exist "outside" the universe
because there is no such place. In quantum cosmology the notion of
an observer becomes, to put it mildly, problematic.

It was against such a background that, in the spring of 1977,
Wheeler and DeWitt invited Everett to a seminar in Austin. There,
during a lunchtime gathering, Everett met a young British postdoc
named David Deutsch. They conversed at length on the 1957
paper, and with encouragement from DeWitt, Deutsch, too, was
persuaded to accept physical reality as a very, very large number of
spacetimes—a *multiverse*.

Many physicists working in quantum mechanics studiously avoid
the whole business of interpretation, a strategy that many—Visser
among them—terms "shut up and calculate." Deutsch prefers the
more respectable-sounding "pragmatic instrumentalism," but he
considers it a rather onerous sort of thinking because it separates
experiment from larger questions, addressing the hows but neglect-
ing the whys. He believes that such willed disregard for the interpretation or meaning of a theory is likely to impede scientific progress.
In a 1980s BBC interview, Deutsch offered a cautionary example.
Pope Urban VIII allowed Galileo to treat the Copernican theory that
the planets revolve around the Sun not as Galileo wished to treat

it—that is, as a description of an objective reality—but only as a mathematical hypothesis useful for predicting the positions of points of light in the sky. Had Galileo's successors followed suit, Deutsch said, Newton's theory of universal gravitation would never have come about.[7]

Accordingly, Deutsch believes that the theory of quantum mechanics should not be divided from its interpretation. In his own speech and writing, he seldom discriminates them. In fact, when he speaks of quantum mechanics he is implicitly evoking the *Everett-DeWitt* (or *"many universes"*) *interpretation*—with a small modification. Everett and DeWitt spoke of "branching" universes, implying that when a universe is confronted with a quantum alternative, it splits or branches into universes in which each alternative is realized. Deutsch prefers to imagine that, instead, there are a great many universes—perhaps an infinite number, but a number that remains constant over time. Originally these universes were identical. But when a quantum alternative presents itself, they partition themselves into groups in which each alternative is realized.

Deutsch's version differs in another way as well. Universes may not only be partitioned; they may be *"un*partitioned," or made identical again. In the case of the double-slit experiment, the two universes existed only as the electron was traveling through the apparatus. When we observed interference, we observed the effect of the two universes becoming identical again. In Deutsch's view, the multiverse is the set of all universes evolving together, and (to borrow his own metaphor) the individual universes are like cogwheels in a great machine. Through interference effects, some are touching directly, and it is impossible for one to move without moving others.

The term "multiverse" does not imply, as some science fiction might have it, that everything imaginable happens in a universe. Rather, phenomena in these universes are restricted by the same laws of physics that govern our universe. Phenomena within them cannot be unphysical, but they can be—and here Deutsch borrows a term from philosophy—*counterfactual*. What exactly is meant by this? To take an example, Thomas Roman once told me that, if he had not observed that the Morris-Thorne traversable wormhole would present a challenge to ideas of causality, someone else would have. Of course, Roman *did* make the observation, and someone else did not. In the context of classical spacetime, his statement refers to an unreal and imagined situation. In the context of the multiverse, however, his statement would refer directly to physical realities—that is, to variants that occurred in other universes. It would mean that, "in most universes in which he (Roman) did not make the observation, someone else did." Moreover, the word "most" in that statement would have real quantitative meaning, because the multiverse itself obeys laws of quantum probabilities, and universes exist in definite proportions.[8]

This means that, in any fundamental sense, there is no difference between a moment in another universe and a moment in another time in our universe. What we are accustomed to thinking of as past moments in our own universe exist in universes that are nearby and that have most influenced our universe. What we are accustomed to thinking of as counterfactuals are moments in universes that are far more distant and have had far less influence on our universe. What we call past moments and what we call counterfactuals, then, are different only in degree of proximity and influence, not in kind.

Physicists take it as a given that no experiment can show one interpretation of quantum mechanics as more valid than another. So it is noteworthy that Deutsch has two ideas for testing the Everett-DeWitt interpretation. One—perhaps the more realistic—involves a computer sensitive to interference phenomena at the quantum level, programmed to record its experience of an experiment for which the Copenhagen interpretation predicts a "collapse" and the Everett-DeWitt interpretation predicts the interaction of parallel universes.[9] Deutsch's other idea is nearer our interests here. As the Everett-DeWitt interpretation provides a novel understanding of time, so an experiment *with* time, Deutsch believes, would prove the interpretation. Whether this experiment could actually be performed is another question. It would, as we will see, require closed timelike curves.

Recall that the 1990 paper by the group of seven physicists that Thorne called the consortium had applied Novikov's self-consistency principle to the grandfather paradox and concluded that pastward time travel need not violate causality. Deutsch regarded this result as dubious for much the same reason that Visser did: the constraints on the time traveler intent upon altering history seemed overly contrived and improbable. Deutsch thought self-consistency as applied to the bootstrap paradox (and the jinn it might produce) just as misguided, but for another reason: it violated a basic principle of the philosophy of science—namely, that complex entities (like life-forms, solutions to problems, and works of art) do not spring into being fully formed. Rather, they emerge from simple states and gradually develop into complex states. In November 1991 he published an article in *Physical Review D* that registered these objections. Its larger point, though, was that both sorts of paradoxes are chimeras, appearing only when we think of time machines in classical terms. If we think of time machines in the context of the Everett multiverse, both sorts of paradoxes vanish.

Time Travel in the Multiverse

Let us assume that the Everett multiverse is real and imagine what might happen if we attempt to force a grandfather paradox. Suppose that Gassendi has received a very large research grant—his largest yet—and is in possession of a time machine. It is Monday, and his plan is this: On Tuesday he will enter the machine, travel pastward, and emerge from the machine on Monday. He will do so unless, on Monday, he sees a version of himself emerge from the machine. If he sees a version of himself emerge from the machine on Monday, he will not enter the machine on Tuesday. Novikov would say that Gassendi has no chance of completing his plan. It would create a paradox, and self-consistency demands that something—a slip on a wet floor or a change of mind—will prevent it. The Everett-DeWitt interpretation requires no such artifice. It does, however, require two universes.

In universe one, Gassendi emerges from the time machine on Monday and joins the version of himself that was waiting. The waiting Gassendi follows through with his plan and does not enter the time machine on Tuesday. Instead, he invites the version that emerged from the time machine to tea. In universe two, no one emerges from the time machine on Monday, so Gassendi follows through with his plan and enters the time machine on Tuesday. He emerges from the time machine on Monday in universe one, to be greeted by a version of himself who invites him to tea. What has happened? In attempting to force a paradox, Gassendi has removed himself entirely from universe two and doubled his presence in universe one.

Gassendi's actions—much like Novikov's jinn—would seem to violate the law of conservation of energy. Recall that the law states that energy cannot be created or destroyed, and that the total amount of energy in the universe remains constant over time. Yet Gassendi seems to have violated the law going in both directions, effectively ceasing to exist in one universe and appearing, quite from nothing, in another. Deutsch would respond that, in his version of physical reality, the law remains perfectly intact. It has simply been broadened: no single universe is a closed system, but the multiverse as a whole is. Although versions of Gassendi may be said to have been destroyed and created in the contexts of individual universes, in the context of the multiverse the number of Gassendis remained constant.

Leaving Home without Leaving Home

Deutsch believes that, if a time machine were possible, it would be a gateway between universes and would permit a number of remarkable circumstances. One of these, which he calls "asymmetric separation," is a relationship between two entities that is experienced as a separation by only one. Suppose Gassendi enacts the plan described in the previous section, and his assistant watches. On Monday in universe one, the assistant, standing beside Gassendi, sees another version of Gassendi emerge from the time machine. The three of them have tea. On Monday in universe two, the assistant, standing beside Gassendi, sees no one emerge from the time machine, so on Tuesday Gassendi follows through with his plan and enters the time machine while the assistant watches. The assistant in universe two waits and waits, but his master never returns. The consequence is that Gassendi's assistant, remaining in universe two, experiences a separation from his master, but Gassendi, having traveled to universe one, does not experience a separation from his assistant. Because the multiverse obeys laws of probability, one could predict the "subjective probability" of the result of such an experiment. From the assistant's point of view there was a fifty-fifty chance that he would not see his master again. But from Gassendi's point of view there was *no* chance he would not see his assistant again.

Deutsch notes that the effect of asymmetric separation might be directed to ensure desired outcomes. If, for instance, Gassendi's assistant has grown tired of his master, he might arrange to escape him altogether. Suppose Gassendi devises the same plan—that is, if he sees a version of himself emerge from the time machine on Monday, he will not enter the machine on Tuesday. His assistant, meanwhile, resolves that if Gassendi emerges from the time machine on Monday, he (the assistant) will enter the time machine on Tuesday. If he does, he is guaranteed to emerge in a universe inhabited by another version of himself and a version of Gassendi who, as long as he follows through with his plan, will enter the time machine on Tuesday, leaving two versions of the assistant "alone" in a Gassendi-less universe.

As we can see, the Everett-DeWitt multiverse would allow travel to a past, but not the past we experienced. If David Lewis's "Tim" travels pastward and kills his grandfather, his action would occur (and *could* only occur) in a universe in which he killed his grandfather. In Everett's original interpretation, at the moment the act is undertaken a universe would split, producing a new "branch" universe in which Tim's grandfather is dead and in which Tim will never be born, and so has no past history in that universe prior to the moment of the murder. In Deutsch's version, at the moment the act is undertaken existing universes partition themselves into a group in which Tim did not kill his grandfather and a group in which he did kill his grandfather. In both the original interpretation and Deutsch's version, the universe we began with would go on as before, with a grandfather who lived to sire a son who, in turn, sired Tim.

Jinn in the Multiverse

Let us continue to assume that the Everett multiverse is real and imagine what might happen if we attempt to force the other sort of paradox—that is, a bootstrap paradox—and so produce a jinn. Recall that what Lossev and Novikov call jinn is something created from nothing. Deutsch, on the other hand would say that what they call jinn, like every other complex entity, emerged from a simple state and developed over time. What makes it different from all other such entities is that it did its emerging and developing in another universe. If I travel into the future and find a completed book manuscript on the kitchen counter, I will have traveled into a universe in which a version of myself *did* labor for months over said manuscript. If I retrieve that manuscript and return to my past, I will have transported it from the universe in which it was created to a universe in which it was not created.

Likewise, Deutsch would say that Lossev and Novikov's clever spacecraft design was the product of considerable labor by engineers, and the spacecraft's memory of the locations of the wormhole mouths was derived from the discoveries of astronomers and astrophysicists—all in another universe. Obviously, for the manuscript and clever spacecraft to travel between universes, there must be some "mirroring" between those universes. As Deutsch writes, "If I am going to try to enact a 'paradox' . . . my universe must become connected with another one in which a copy of me has the same intention as I do, but by carrying out that intention ends up behaving differently from me."[10]

Deutsch disavowed the creation of knowledge via a bootstrap paradox (jinn of the second kind), but he suggested that the structure of the multiverse might allow knowledge to appear from nothing in a given universe.

Presented with a quantum alternative, two identical universes will, over time, grow farther apart and more different, and at some point in the divergence, one universe will contain knowledge not contained in the other. If a civilization in the universe possessing such knowledge had access to a time machine, it might send that knowledge pastward to a civilization in the other universe. Because such transmissions could occur only between nearby universes, their number, although large, would be finite. Nonetheless, because there are no conservation laws regarding knowledge, the total amount of knowledge across universes would be increased.

Knowledge given us by time travelers from the future would not be knowledge of *our* future, but of a future much like it. Knowledge of an undesirable future might tell us how to avoid futures like it; knowledge of a desirable future might tell us how to direct ourselves toward futures like it. The value of either type of knowledge would be incalculable.

Free Will in the Multiverse

What of free will in the multiverse? Although the entire history of
the multiverse is fixed, from the perspective of someone at a given
moment in any universe, only the past is fixed. Recall that the
determinism of the nineteenth century and the block universe in
the twentieth century deny us—or seem to deny us—free will.
Physicists who delved into philosophy suggested that the quantum
universe or, more precisely, the indeterminate nature of the quantum universe, restored it. One of these was Heinz Pagels, who, in a
memorable turn on Einstein, wrote, "The God that plays dice has
set us free."[11]

Deutsch feels rather differently. The indeterminacy and randomness associated with most interpretations of quantum mechanics
may make the future of a given subatomic particle unpredictable;
and they might by extension make all our futures unpredictable, as
we are, after all, composed of subatomic particles. But unpre-
dictability does not in any meaningful sense endow us with free
will; that is, it does not provide us a means to determine our future.
In this sense it is no improvement over determinism or a block universe. A multiverse, however, *would* allow free will, if not in the
aggregate, then at least subjectively. Each universe within the multiverse is deterministic, but I exercise free will by moving among
them. In other words, all possible futures are there. When I make a
choice, I am merely deciding which one I want to be in.

CHAPTER ELEVEN

EXPLAINING THE APPARENT ABSENCE OF TIME TRAVELERS

IN WHICH THE READER IS INVITED TO THINK IN THE MANNER
OF BEINGS FAR MORE POWERFUL AND WISE

Where are they?
— Enrico Fermi

Deutsch's colleagues were aware that his ideas regarding time travel assumed an interpretation of quantum mechanics and physical reality that they did not share, and this may have been the reason why there were few responses to his paper. Matt Visser noted that Deutsch did not so much show that closed timelike curves would offer paths to the other universes as assume that they did.[1] At least one physicist, however, was sympathetic to the Everett-DeWitt interpretation, and he has already appeared in our story. It was Allen Everett, who had shown that the Alcubierre "warp drive" could, in principle, be made a time machine. Everett shares a surname with the man who first suggested the multiverse interpretation, and although not related by blood, he was an undergraduate at Princeton when the other Everett was working under Wheeler, and thinks that they once played bridge together.[2]

Allen Everett read Deutsch's article with great interest and found it flawed. In a 2004 piece, he argued that Deutsch had failed to account for the resolution time of the device detecting objects emerging from the time machine, and he suggested that macroscopic objects (including human beings) would be broken into microscopic fragments, with different fragments winding up in different universes.[3]

Everett's paper was one of only a few works on closed timelike curves published that year; the torrent of the 1990s had by then diminished to a small stream. Everett himself did not attribute the falling off so much to the 1997 work of Kay, Radzikowski, and Wald, as to the realization that all avenues had reached impasses. He said, "It's not quite clear where one goes from where we are."[4]

In those same years, however, the lay public had shown greater interest in wormholes and causality violation, both of which were much in evidence in television and Hollywood films. In late 1996 the BBC's *Horizon* series aired a program on the subject of closed timelike curves.[5] It featured many of the figures who played large roles in the inquiry: Thorne, Novikov, Hawking, Visser, Deutsch, and Wheeler. It also featured Carl Sagan. Closed timelike curves were not among Sagan's fields of expertise, but at the time he was the first person whom the press thought to consult on any aspect of science, and despite his advancing illness (a rare disease called myelodysplasia), he obliged. In fact, he agreed to a separate interview.[6]

As might be expected, Sagan's performance was masterly. In a fashion that the public recognized as uniquely his, he evoked wonder and awe and alluded to the practice of science as serious play. He managed to distill the puzzles posed by temporal paradoxes into a few clean phrases. When asked about his own contribution, he said, "I find it marvellous, I mean literally marvellous, full of marvel, that this innocent inquiry in the context of writing a science-fiction novel has sparked a whole field of physics and dozens of scientific papers by some of the best physicists in the world."

During the interview a question arose that we have yet to

address: Does not the fact that we have no evidence for time travelers from our future strongly suggest that time travel is impossible? Sagan argued otherwise, enumerating many credible reasons to explain the lack of evidence. That he was able to do so easily may be owed to the fact that the reasons were much the same as his reasons for why we have no evidence of extraterrestrials. It was logical borrowing, given that both the extraterrestrials and time travelers were members of "sufficiently advanced civilizations"; that the second group would be our descendants made little difference.

In 1979 the "extraterrestrial question" was the centerpiece of a conference held at the University of Maryland in College Park.[7] It had occurred to many that, although the distances between the stars were vast, the galaxy was astonishingly old, and there were Sun-like stars five billion years older than the Sun. By inference, there was the possibility of Earth-like planets of the same age. This meant that there had already been ample time for a starfaring civilization—even a slowly moving one—to have colonized the galaxy many times over. It was a startlingly simple proposition: if there were extraterrestrial beings, there would be ample evidence. And it was an equally startling conclusion: because we do not see even one, there are none.[8]

At the Maryland conference, though, participants voiced doubts. Some said that perhaps societies grew disinterested in colonization; perhaps they destroyed themselves or they underwent technological decline. Others, admitting that it seemed unreasonable to apply such social theories to *all* of the million societies whose existence was being posited, suggested more universal impediments—that interstellar travel was difficult and expensive. Still others suggested that perhaps we live in a relatively unpopulated region, the "Sahara Desert" of the galaxy. Harvard's John Ball posited that the galactic community has, for one reason or another, consigned us to a kind of quarantine.[9]

Because the question is in many ways congruent with the question that the BBC posed to Sagan, we may allow ourselves to borrow a few of its suppositions, and (with modifications) a few of its answers. First, though, we would do well to clean out some cobwebs. Science fiction authors have imagined any number of physical phenomena that might prevent us from seeing travelers from our future: a side effect of time travel renders them invisible; they are able to stay in our time for an arbitrarily short period; their manifestation is indistinct or otherwise imperfect, so they are visible or audible to us as phantasms or poltergeists; or the structure of spacetime itself disallows time travel. All these premises exist to serve a dramatic need, and they have little or no basis in the real (albeit theoretical) physics discussed earlier. We will stay clear of them here.

The previous chapter described David Deutsch's ideas on how time travel would be manifest in the Everett-DeWitt multiverse. To the question of the absence of evidence, Deutsch said that, although time travel may be possible and may be undertaken, spacetime is structured in such a way that we may never see its effects. If there *are* pastward time travelers, they may simply be in other universes.

He had a second, still more provocative answer as well. Recall the experiment in which Gassendi's assistant managed to separate himself from Gassendi, and imagine the same effect carried out on a very large scale. If our descendants discovered the mouth of a wormhole time machine large enough to accommodate the Solar System and had the means to move the Solar System into it, they would be able to create as many copies of the Solar System as they liked, all in different universes. For the Solar System and its inhabitants, the endeavor would carry no risk. There would be a fifty-fifty chance that they would disappear from this universe, but no chance that they would not appear in another.

Moreover, Deutsch says, "If the effect is typically used by civilizations who are able to, and the time taken to reach the required technological level is small compared with the time taken for a civilization to spread across the galaxy, then it would be very unlikely that any young civilization such as ourselves would yet have observed other intelligent life even if civilizations come into existence quite frequently in the galaxy."[10] Why might we (or any other civilization) wish to move into another universe? In part, to gain resources of matter and energy. Deutsch imagines that if two civilizations have competing plans for the galaxy's resources, then by using the closed timelike curve as Gassendi's assistant used it, each might manage to stay out of the other's way.

Let us put Deutsch aside for the moment, however, and suppose that we inhabit the four-dimensional Hausdorff differentiable manifold discussed in Chapter 9—that is, the traditional model of spacetime assumed by most of the time machine models that we have described. In this physical reality, any imaginable time machine—a Morris-Thorne wormhole, Gott's cosmic strings, an Alcubierre space warp, or a Krasnikov tube—would allow the time traveler to venture pastward only to the moment of the machine's construction.[11]

If, for instance, the first time machine is created on January 1, 2100, then that date would also mark the earliest moment its user could reach. Subsequent times may be overrun with transtemporal tourists, but none of them could ever visit us in the early twenty-first century, nor could they visit anyone in the sixteenth century or the first. All the ideas of pastward time travel that we have examined require time travelers visiting us to use time machines that exist in our present. This means that, unless someone has secretly developed a time machine—and as we have seen, the energies required make such an eventuality unlikely—then to visit us, time travelers from our future will have to use natural time machines, or time machines constructed by extraterrestrial civilizations long ago. Suppose they exist. Can we imagine reasons that we have no evidence of their use?

In fact, we might imagine seven:

1. Time travel requires the use of natural or preexisting wormholes that are never discovered.

Closed timelike curves exist somewhere in the universe but are not found. As suggested earlier, they may be very short-lived, extremely rare, or far beyond the range of our telescopes.

2. Time travel proves to be unacceptably expensive or dangerous.

The suitably advanced civilization that has been under discussion is bound only by the laws of physics. We will return to that civilization shortly, but we might pause for a moment of realism and admit the possibility of more mundane impediments.

It may be that closed timelike curves are discovered but they are at distances so great that travel to them by normal space is prohibitively expensive. If, on the other hand, time travel is shown to be economically viable, it may be deemed not worth the risk to life. Perhaps the suitably advanced civilization attempts it, there is an accident, and the effort is ended. Such an eventuality would have precedent. After the first manned lunar landings in the late 1960s and early 1970s, space programs in the United States were scaled back for a set of reasons: The cold war, which had motivated the space race, had ended, and the political necessity (whether perceived or real) ended with it. Arguments that lunar landings were too costly had more weight. According to some, there arose new concerns for safety.

3. Time travel ceases to be interesting.

Kip Thorne moved away from the subject in the early 1990s to pursue research in gravitational waves. Likewise, perhaps, other scientists will also grow less interested in time machines and direct their attentions elsewhere. Or perhaps there will arise a large-scale shift in the nature of scientific culture, as in, for example, a movement away from experimental science to more purely philosophical approaches.

We are almost inured to technological progress, and it is easy to believe that such progress will continue unabated. But it is possible that, in the long view of human history, such progress is a short-lived aberration. Humans have existed for roughly two hundred thousand years, but scientific and engineering progress has been a hallmark of only the last six thousand, and even in that span it proceeded with fits and starts. Long intervals, like the Middle Ages in western Europe, saw relatively little technological development. Although scientific reasoning had antecedents in several cultures, what most would call formal scientific method took hold only in the seventeenth and eighteenth centuries, and even then only in western Europe. It may not be as robust as we would like to imagine.

Suppose, then, that for any of these reasons, or a combination of them, only a few time travelers ever undertake the journey. Some travel to times later than our own where their presence is much publicized or where they pass unseen. Some journey to our time and times previous and are, for various reasons, unnoticed. After these few voyages, time travel ends forever.

4. Time travel is possible, but prohibited.

Anthropologists have well-founded concerns about contaminating other cultures. In encounters between societies of radically different technological levels, as was the case when a few hundred Spanish conquistadors encountered the Aztec, the less technically advanced suffers. Even when the advanced society has good intentions, the other is likely to be harmed by sheer exposure. Much high-minded science fiction presents the situation as a dramatic moral dilemma, but the concern has at least one real historical cognate. In 1961 the Brookings Institution began a NASA-sponsored study to identify long-range goals of the U.S. space program. The resulting report, which discussed the implications of the discovery of extraterrestrial life, included this rather sobering note: "Anthropological files contain many examples of societies, sure of their place in the universe, which have disintegrated when they had to associate with previously unfamiliar societies espousing different ideas and different life ways; others that survived such an experience usually did so by paying the price of changes in values and attitudes and behavior."[12]

Perhaps, failing Hawking's chronology protection conjecture, there arises a widely accepted ethical prohibition against time travel or a law prohibiting it.[13] Suppose that the civilization capable of time travel has ethics that have matured along with its science, and in order to protect the inhabitants of the past, or to protect the past itself, it has established prohibitions against time travel. Even our own (relatively primitive) civilization has occasionally refrained from exerting its power, setting aside tracts of land as wildlife refuges or wilderness areas in which species may live free of our interference. An animal held in a "perfect" refuge is unaware that it is in a refuge, and is insensible of the presence of game wardens and sightseers. Perhaps we are accessible to time travelers but are protected in a perfect refuge.

Alternatively, future generations may regard us as ethically undernourished and dangerous, and place us under an enforced isolation in the interest of protecting themselves.

5. Time travelers are careful to hide their presence.

Time travelers may remain inconspicuous by any number of strategies that would violate no natural laws. Perhaps they are observing from some distance in space, or watching through robot proxies that somehow make themselves invisible to our telescopes and radar. Perhaps they are much nearer but routinely drug or hypnotize witnesses. Paul Davies (a physicist, not a science fiction writer) has suggested that very advanced civilizations, to conserve energy and make travel more efficient, will reduce themselves in size.[14] Very small time travelers could be quite discreet. Or perhaps they are simply very well disguised. They may be living among us now, dressed in period clothing (that is, *our* period) and carefully schooled in the customs and language of our time.

Exactly who are they? Science fiction writers have offered many answers and might here be welcomed back to the discussion. Perhaps they are the twenty-fifth-century equivalent of historians or anthropologists. Perhaps they are a particularly adventurous sort of tourist. Some visit for a few weeks or months and then return to their own time, others prefer to stay longer, and a few remain for the duration of their lives. Occasionally these temporal expatriates betray their time of origin, letting slip their knowledge of a future event or technology. We judge them eccentric, brilliant, or insane.

6. Civilization does not survive long enough to develop time travel.

In 1960, early in the search for extraterrestrial intelligence, radio astronomer Frank Drake devised an equation by which the number of civilizations in the galaxy might be estimated. All the equation's unknowns—number of Sun-like stars, proportion of habitable planets, proportion of habitable planets supporting life, and so on—were factors in the strict mathematical sense: if any one of them were zero, the number of civilizations would also be zero. As it happened, many of the factors canceled themselves out or were equal to one, so the result depended entirely upon the last unknown in the series—which happened to be the longevity of a communicative civilization.

In 1960, this particular number was of some immediate concern because there seemed to be a fair chance that the United States and the Soviet Union were sliding toward a nuclear confrontation. Drake and several of his colleagues believed that, although a civilization's nuclear age was the most dangerous time in its existence, it was short-lived, and a society that managed to survive it could anticipate a long life. For this reason, Drake expected the lifetimes of civilizations to be very short or very long.

Others were less optimistic. The Soviet physicists who had been deeply involved with nuclear weapons research had developed a decidedly dark view of humanity's prospects. Iosef Shklovskii, known as the father of SETI in the Soviet Union, noted that humanity learned that it was possible to send radio signals between the stars only in the 1930s; a mere two decades later it had acquired enough thermonuclear weapons to destroy itself many times over. He doubted that any civilization would long survive its technological stage.

If we are seeking a cause for human extinction, we do not need nuclear annihilation. Numerous scenarios might accomplish the same end—a large asteroid impact or a plague (natural or artificial), to name but two; and these might be more probable than we care to suppose. In his book *Our Final Hour*, Martin Rees tallies an impressive number of possible apocalypses and gives humanity a fifty-fifty chance of surviving into the next century.

Thus we borrow from such thoughts to confront a particularly unpleasant possibility. Before a natural or preexisting time machine is discovered, civilization destroys itself or is destroyed.

7. Time travelers prefer to travel to times other than our own.

There is a final possibility, one that we might find the most unflattering of all, as it cuts to the heart of our self-esteem. Perhaps the reason we have seen no time travelers is that we are simply of little interest.

As of this writing, the best estimate of the age of the universe is that supplied by the Wilkinson Microwave Anisotropy Probe, which puts it at 13.7 billion years, plus or minus 1 percent. When we shrink that span—from the big bang to the present—to a single year, we may be surprised at how little of it concerns us directly. The Sun and its planets do not form until mid September. Simple life-forms appear on Earth in early October. A month later the atmosphere begins oxidizing. On Christmas Eve the first fish crawl onto land, and a few days later the dinosaurs appear. On December 30

AN ARTIFICIAL TIME MACHINE

What, then, of an artificial time machine? If we assume no chronology protection mechanism, then the major impediment to its construction (at least the one we know of) is the energy required. Visser had calculated that a wormhole with a throat radius of one meter required negative energy of a magnitude equivalent to the total energy generated by ten billion stars in one year. So to undertake pastward time travel the sufficiently advanced civilization will need to harness the energy of one-tenth of the stars in the galaxy. We have implied that such a scenario is unrealistic. But is it? Perhaps surprisingly, the question was addressed by a well-known piece of informed speculation.

In 1964, Russian astrophysicist Nikolai Kardashev authored a paper suggesting that extraterrestrial civilizations might generate enormous amounts of waste radiation that would be detectable over great distances as infrared wavelengths. In the 1960s, the Soviet Union dominated the search for extraterrestrial intelligence and frequently adopted bold strategies. Rather than searching the vicinities of nearby stars, the Soviets used nearly omnidirectional antennae to observe large portions of sky, counting on the existence of at least a few supercivilizations capable of radiating enormous amounts of energy.

Kardashev contended that, over time, a technological civilization would continue to increase its energy use, not merely linearly but exponentially, and that civilizations might be categorized by the amount of energy they controlled. What he termed a "Type I Civilization" used all the solar energy falling on its planet and all the energy within it—that is, solar, geothermal, wind, and tidal. A "Type II Civilization" would use all the energy of its star, perhaps by surrounding it with a system of orbiting structures designed to intercept and collect it. A "Type III Civilization" would use the energy of an entire galaxy, perhaps enclosing all its stars in such structures, or by a means that we have not imagined.

Kardashev wrote that, 3200 years from the present, "the energy consumption per second will be equal to the output of the sun per second, [and] in 5800 years the energy consumption will equal the output of 10^{11} stars like the sun."[a] That number—one hundred billion—is roughly the number of stars in the galaxy. Obviously, the difference between Types II and III is great, and finer discriminations have been proposed. Still, it seems that we will need to reach a stage between Type II and Type III before we can expect to control the energies necessary to build a time machine.

Naturally we might wonder whether humanity will survive that long. In 1993, J. Richard Gott introduced a formula by which one might predict the life span of most anything—a technology, a political institution, or a Broadway musical.[b] He based his idea on the straightforward assumption that, when we observe something, unless we have knowledge to the contrary we are probably observing it at no privileged moment in its history. Mathematically, there is a 95 percent probability that we witness a given phenomenon during the middle 95 percent of its existence. (The formula would work as well for other numbers, but Gott chose 95 percent because scientists traditionally use it as a standard for predictions.) Generally, the results would seem to yield few new insights, but in certain cases they have rather interesting implications.

For its January 1, 2000, issue the *Wall Street Journal* asked Gott to estimate the life spans of a range of physical structures and institutions. He predicted with 95 percent certainty that the Great Wall of China would survive between 56 and 86,150 years into the future, the Internet between nine months and 1209 years. Most intriguingly for our purposes, he predicted with 95 percent certainty that the human species will survive until at least 5100, and not beyond 7.8 million, years from the present. The lower threshold for humanity's survival, when placed against Kardashev's model, yields an interesting date. It is only seven hundred years before the advent of a Type III Civilization, and nearly two thousand years into the lifetime of a Type II Civilization. From this we might conclude that, if nature allows time travel and all the other impediments discussed here are met and overcome, the first pastward time travelers will be members of a late Type II Civilization.

[a] Kardashev, "Transmission of Information," 218.
[b] Gott, "Implications."

arise the first small warm-blooded animals. Early on New Year's Eve *Homo erectus* appears and disappears, and Neanderthal comes into view and vanishes. The rise and fall of civilizations; the vast migrations of peoples; the discoveries of continents; the struggles of nations and empires; the creation of art, philosophy, and religion; the development of science and technology; all that has been accomplished by human civilizations—in fact, everything that we call recorded history—appear only in the final ten seconds of the year, roughly the time it took to read this sentence.

That, at least, is the typical presentation of what is called the "cosmic clock." It is somewhat misleading because, by equating the last moment of the cosmic year with our present, it suggests that this moment is somehow the culmination not merely of life on Earth (a claim that most evolutionists would regard as misunderstanding the mechanism of evolution), but of everything that has occurred in all 13.7 billion years of the universe's history. We might achieve a more objective perspective if we let the clock continue to run for, say, another three billion years past this moment and imagine the perspective of our descendants, members of a supercivilization living at that time.

Three billion years is a very long span. There would be almost as much time between that supercivilization and us as there is between us and the single-celled organisms that are the oldest life-forms known to exist on Earth. We might reasonably suppose that members of that civilization would have evolved well beyond recognition. Like the alien civilizations of much science fiction, they would regard us as objects of study and would be no more inclined to communicate with us than we would be inclined to communicate with a bacterium, even if we could be convinced it was our ancestor. But to be uninterested in us they would not need to be scientifically and technologically advanced. A carpenter ant is not much interested in a snail either. They would need only to be sufficiently *different*, and three billion years is enough time to develop a great measure of difference.

Suppose that members of this supercivilization discover a preexisting wormhole time machine (naturally occurring or constructed by another civilization) at a date that, on the cosmic clock, would be April 1 of the new year. And suppose that the time machine allows access to the past, all the way back to the very early universe. With most of a year and three months of the cosmic clock to explore, they might not regard ten seconds late on December 31 as particularly noteworthy; indeed, they might overlook those ten seconds altogether.

Of course, we may counter that the brevity of our species' existence thus far in no way implies that we are unimportant or uninteresting; it implies merely that we are young. Fair enough. We may also suggest that, as far as we know, we are the most complex arrangement of atoms and molecules in the universe, a rare example of a place where entropy is locally (and only momentarily) reversed. It is even possible that, in the words of some cosmologists, we are the universe first becoming conscious of itself. But we are also likely to be prejudiced. The young believe themselves both important and interesting. It may well be that we think humans are interesting because we *are* humans.

There may be no reason to assume that we will be the only intelligent species to inhabit Earth. In fact, if we are prepared to be wildly mistaken (always a prudent attitude in matters like these), we might hazard a guess as to the chance that Earth might give rise to another such species. How great a chance? We would need first to establish the span during which Earth might be suitable for life. Even assuming that none of the catastrophes alluded to earlier come to pass, if the Sun continues to warm at its present rate, this planet may have about one billion years of habitability remaining. How much time does nature need to evolve an intelligent species? Our own unprepossessing ancestors—small primates resembling tree squirrels—were given their opportunity with the sudden extinction of the dinosaurs, some sixty-five million years ago. There are fifteen such spans in one billion years. If this represents an average (an admittedly big "if"), then there are fifteen more chances for the evolution of intelligent life on Earth. Suppose, then, that time travel is first invented by a species that appears after the demise of our own. They might have even less reason to regard us as interesting than would our own descendants.

CHAPTER TWELVE

TIME MACHINES AT THE ENDS OF TIME

IN WHICH PHYSICISTS DEBATE WHETHER THE UNIVERSE WILL END
IN A COLD AND EMPTY WASTE OR A FIERY IMPLOSION,
AND TIME MACHINES ARE IMAGINED TO HAVE
A NEW AND SURPRISING USE

In the winter of 1930, physicist Arthur Eddington, along with much of Britain's reading public, was captivated by a startlingly original work of fiction. It was a short novel with the peculiar title *Last and First Men*, and its commercial success was unexpected, among other reasons because it had no central character, and in fact very few individual characters at all. Given the novel's scale, such omissions were understandable. *Last and First Men* was the imagined chronicle of the next two billion years of human—and *post*human—history. It was also, quite probably, the first instance of an author using known science to imagine in detail something like what we have been calling a "suitably advanced civilization."

The creator of this curious work was an equally curious man, a forty-four-year-old scholar named W. Olaf Stapledon. Stapledon held no academic post, and his formal training was in philosophy. Yet he was a regular reader of the journal *Nature*, and his attentiveness to developments in astronomy and evolutionary biology allowed him to imagine in detail a span of time in which not mere civilizations, but whole species calling themselves human, arise—in the process adapting to enormous changes in their environment—and fall.

By the book's final chapter the Sun has grown so hot that the inner Solar System is uninhabitable, and the eighteenth human species (which Stapledon terms the "Eighteenth Men") have colonized the planet Neptune. But before long this haven too comes under threat. Astronomers have learned that the swollen Sun will soon erupt with a violent storm that will sweep through the entire Solar System. From such calamity they have no means of escape; their "ether ships" are incapable of interstellar voyages. So it is proposed that biologists engineer a miniaturized human seed, to be cast into space from strategic points in Neptune's orbit and allowed to be carried outward by the solar wind. Some of this seed, they hope, may one day find hospitable ground on a distant planet, and so accord their species a modest sort of survival.

This plan, however, also meets with difficulties. The prospect of their own demise has instilled a specieswide despair, and the Eighteenth Men cannot summon the will to complete the work. So they undertake a second project—one that will call upon a feature in their rather highly evolved neurophysiology. The Eighteenth Men can intuit spacetime directly. Moreover, by what Stapledon terms "a partial awakening, as it were, into eternity,"[1] they have taught themselves to influence past minds in such a way that they can, to some degree, direct their own history.[2] Now, however, they hope only to inhabit those minds long enough to regain their ancestors' passion for life and thereby be invigorated to complete work on the seeding project. It is on this poignant and uncertain note that the novel ends.

Stapledon's vision of a long-lived humanity, appearing at a time when fascism was spreading across Europe, served as a kind of spiritual tonic. Several of Stapledon's contemporaries, inspired by utopian impulses, also attempted to envision a far future. In 1923 the geneticist J. B. S. Haldane produced a paper called *Daedalus: or, Science and the Future*; and in 1929, John Desmond Bernal published a monograph called *The World, the Flesh and the Devil: An*

Enquiry into the Future of the Three Enemies of the Rational Soul. Also about this time, Jesuit priest and philosopher Pierre Teilhard de Chardin was developing his own account of the long unfolding of the material cosmos, an unscientific (albeit quite poetic) description of the long ascendancy of life.

The first half of the twentieth century had seen predictions of the future of humanity; the second half had seen—in a 1977 piece by physicist Jamal Islam—a prediction of the future of the physical universe. It was for a physicist with a philosophical bent to pull these strands together. In 1979, Freeman Dyson, of the Institute for Advanced Study, published "Time without End: Physics and Biology in an Open Universe." It proposed a means by which intelligent life might survive not for a mere two billion years, but for a literal eternity.

Dyson and Life in a Distant Future

Fifty years earlier, most scientists would have considered such a prospect not merely unlikely, but impossible. Astronomers reasoned that since all closed systems eventually reach thermodynamic equilibrium, and since the universe is a closed system, there would come a time when the stars burned themselves out and every part of space settled into the same very low temperature. Because life depends upon warmer and colder places and the transfer of heat between them, it could not exist in such an environment. Eddington called this the "heat death" of the universe. In the 1930s he and contemporary astronomer James Jeans introduced the concept to the lay public, which (perhaps to the bemusement of our own relatively shortsighted age) regarded it with no small anxiety.

The sentiment was best expressed by philosopher Bertrand Russell, writing, "All the labors of the ages, all the devotion, all the inspiration, all the noonday brightness of human genius, are destined to extinction. . . . The whole temple of Man's achievement must inevitably be buried beneath the debris of a universe in ruins."[3] Such may yet come to pass, but the premises of the heat death have been called into question. In the late 1920s, astronomers had confirmed that galaxies in all directions were moving away from our own, and that the universe—and space itself—were expanding. By the middle of the following decade some had interpreted this to mean that maximum possible entropy was always increasing faster than actual entropy. As long as the universe continued to expand, it would never reach thermal equilibrium.

The distance we can see into the universe is limited by several factors, one of which is the speed of light. Because the universe began 13.7 billion years ago, it follows that we can see in no direction farther than the distance that light can have traveled since that time—that is, no farther than 13.7 billion light-years. The expansion of space makes the actual distance visible considerably greater, but in any case, as the universe ages light is given more time to travel, so astronomers of the future will be able to see farther than can astronomers at present.

We might imagine the universe we see (the *observable universe*) as the volume within a bubble at whose centerpoint we are, and whose outer surface (the *cosmic horizon*) is expanding. Outside the cosmic horizon is a larger bubble—the entire universe—and it, too, is expanding. This expansion means that the average density of matter in a given volume of space is diminishing, and all galaxies (both those inside the cosmic horizon and those beyond it) are growing farther apart.

In 1979, Dyson held with evidence that the observable universe was expanding at a faster rate than was the entire universe. The relatively faster expansion of the observable universe meant that, even though all galaxies were receding from us, just inside the cosmic horizon more and more of them were coming into view. Although the universe that Dyson imagined was growing emptier for a given volume of space, its *total* volume was increasing, and it followed that human prospects were limitless. "No matter how far we go into the future," he wrote, "there will always be new things happening, new information coming in, new worlds to explore, a constantly expanding domain of life, consciousness and memory."[4]

Indeed, it was this "open" universe that would be home to the sort of life that Dyson imagined. Such life would face difficulties. Although the heat death was no longer even a distant threat, there remained the challenges posed by an ever-diminishing supply of usable energy in an ever-colder universe. Dyson proposed a "scaling hypothesis" according to which organisms might adjust to less energetic environments by slowing their metabolisms. Intelligent organisms, he noted, would have their own problem: because thought is a product of metabolic processes, they would be obliged to slow their rates of consciousness. Although Dyson admitted that he could not imagine such organisms in detail (he could not, for instance, know whether there were functional equivalents of muscles or nerves), this did not mean that they were impossible. Most biologists, he noted, would be hard-pressed to imagine a cell of protoplasm, had they never seen one.[5]

Tipler Imagines Another Future

A little more than a decade later, a physicist with a rather different set of suppositions argued that the expansion would cease and the universe would ultimately collapse into itself. More provocatively, he proposed that *before* this end occurred, against all expectation and intuition, there would be enough time for a future that, though finite in the usual sense, was "subjectively eternal" and unsurpassingly glorious; and that by exploiting the energy differentials in a collapsing universe, our descendants would create "an abode which is in all essentials the Judeo-Christian Heaven."[6]

The provocateur in question was Frank Tipler, the self-described physicist with chutzpah, who in 1974 had hypothesized a "rotating cylinder" time machine. It had been twenty years since the time machine paper, and in the interim Tipler had written a great deal on a range of subjects. By 1994 he had been a full professor in the mathematics department at Tulane University for nearly a decade, but no one would say he had mellowed.

The book in which Tipler predicted a human-engineered heaven was *The Physics of Immortality*. As might be expected, its critical reception was mixed. There was awe at the range of its erudition and mild bewilderment as to its inspiration. One reviewer called it "a wonderfully ambitious, painfully sincere tour de force—an attempt, sometimes brilliant, sometimes absurd, to stretch scientific reasoning to its breaking point."[7] Thorne offered qualified praise, saying, "I think big pieces of Frank's argument are brilliant and beautifully done. But it is in the leaky joints between the pieces where I think the problems lie."[8] Many readers were far less kind, and most physicists regarded the book as an unworthy effort from an otherwise much-respected scientist.[9]

To his credit, though, Tipler had offered his idea as a "testable physical theory,"[10] and like all theories it rested upon certain hypotheses. As it happened, one of them would soon be disproved. In the late 1980s and early 1990s, astrophysicists were able to make ever more precise measurements of the mass density of the universe, and when they summed the mass from stars, dust, and gas, and the mysterious stuff called *dark matter*, they arrived at a value far below that necessary to stop or even slow the expansion. By the mid 1990s, most cosmologists were persuaded that a collapse was unlikely, that the universe was *open*. That meant no crunch, no collapse, and no energy differentials to exploit. It also meant that if life were to survive into the far future, it would be in a universe like that described by Dyson.

Then, in 1995, physicists Fred Adams and Greg Laughlin, with benefit of the discoveries of astrophysicists in the sixteen years since Dyson's article, undertook yet another speculation of the future of the universe—this the most ambitious to date.

Current Knowledge of the Fate of the Universe

The first generation of stars formed when the universe was a few million years old. A billion years later the galaxies arranged themselves in larger structures called clusters, and these in still larger structures called superclusters. From here on, time spans under discussion become truly mind-boggling, and to make them manageable we will again resort to scientific notation, keeping in mind that 10^{14} represents a number ten times larger than 10^{13}.

The current era is a relatively active period when stars are forming, living, and dying. It ends when the universe is 10^{14} years of age and all stars have cooled and faded to stellar remnants—white dwarfs, brown dwarfs, neutron stars, and black holes. This begins a tremendously long (and relatively quiet) period; its stillness is interrupted only when two white dwarfs collide and create a supernova explosion that brightens, however briefly, an otherwise darkened galaxy. This period, which lasts until the universe is 10^{28} years of age, would seem utterly inhospitable to life, and perhaps it will be. Nonetheless, taking a page from Dyson, Adams and Laughlin imagined ways in which living organisms, and even intelligence, might survive and flourish.

White-dwarf interiors are unimaginably dense—10^{14} grams per cubic centimeter—but their atmospheres would allow mobility. Those atmospheres contain oxygen and carbon, and although they are quite cold, they are warm enough that the chemicals could interact in interesting ways. Still more compellingly, they will remain stable until the universe is 10^{25} years of age. This would be a hundred billion times as long as it took for life to appear on Earth. Such longevity, Adams and Laughlin suggest, implies that life in white-dwarf atmospheres is not merely possible; it is likely. However, they also note that it would be a life quite unlike our own. In accordance with Dyson's scaling hypothesis, metabolisms and rates of consciousness would be very slow. An intelligent creature living in a white-dwarf atmosphere might take a thousand years to complete a single thought.[11]

Although such beings would be very long-lived, they would not be immortal. When the universe is 10^{40} years old, even protons have evaporated into a diffuse radiation, and the only remaining stellar remnants are black holes. Their radiation provides all the warmth available anywhere, and it is precious little: the surface temperature of a black hole is one ten-millionth of a degree above absolute zero.[12] Eventually, even black holes evaporate, and the larger they are, the more time they take to do it. When the universe is 10^{100} years old, even the most massive black holes are gone, leaving a few electrons and positrons drifting in unimaginably vast reaches of space. This is a time, as Shakespeare seems to have prophesied, "When creeping murmur and the poring dark / Fills the wide vessel of the universe."[13]

In the last decade of the twentieth century, findings in cosmology were coming so quickly that they outpaced the schedules of book production. As Adams and Laughlin worked, cosmologists had compelling evidence that the universe was open, but there remained doubts. In the 1990s, two groups worked independently to gauge the rate of expansion. They measured the redshift of the most distant object whose absolute luminosity we know—a species of star called Type IA supernovae—and took it as characteristic of the expansion rate of the very early universe. They then compared it with the redshift of nearby galaxies, which they assumed typical of the more recent expansion rate, and graphed the rates on a curve. The tentative results, announced by both groups in 1998, surprised everyone. All evidence suggested that the rate of expansion was not slowing; neither was it steady. Contrary to all expectation, it was *accelerating*.

If this acceleration is real (and as this book goes to press, reservations remain), then its cause is unknown. It may be that the vacuum itself exerts a negative pressure of the type featured in ideas of inflation. It may be that the universe is filled with an all-pervasive fluid that has a negative pressure—Michelson and Morley's aether in a more sophisticated form. In any case, an accelerating expansion means a fate for the universe that, if anything, is irredeemably darker and colder than that implied by a slowing or steady expansion. It means that the cosmic horizon can never catch up, that it will not bring new galaxies into view; in fact, as space expands it will take galaxies that are now in view beyond the horizon. Our sky will grow ever darker.

The Final Energy Crisis

In 2002, physicists Lawrence Krauss and Glenn Starkman pondered what an accelerated expansion would mean to prospects for life in general and for the longevity of a suitably advanced civilization in particular. As the universe expands and space is stretched, energy and matter become ever more diluted. Consequently, the suitably advanced civilization would be faced with the energy crisis to end all energy crises. It might try to retrieve matter, but the process would cost more, in terms of energy, than the matter retrieved. It might decide to stay put and allow gravity to do the work—for instance, by constructing a black hole so that matter might be drawn toward it. But because the strength of gravity between two masses falls off in direct proportion to the square of the distance between them, the ongoing dilution of matter means that, over time, gravity itself becomes less and less effective. Other solutions—extracting energy from cosmic strings or the quantum vacuum itself—would be, for various reasons, as unworkable. It seems that the suitably advanced civilization, no matter how clever, would not be able to avoid its own extinction.

New Uses for Time Machines

With sufficient ingenuity and planning, though, the civilization might forestall that extinction. It could do so by subjecting energy or matter to time dilation, thus preserving it for use when such resources have grown scarce. Recall that there are three ways to effect time dilation: linear motion at relativistic velocities, circular motion at relativistic velocities, and/or subjection to a strong gravitational field like that near a black hole or inside a mass shell. Any of them might provide a sort of long-term storage—but all would exact a cost. To accelerate the package to relativistic speeds, the civilization would need considerable energy. It would need more energy to slow the package near the completion of its journey—at a time when, by definition, energy is in even shorter supply. Moreover, although the package would be "coasting" on its outbound and inbound legs, it would be slowed by the expansion of space, which drains even kinetic energy. As if these were not problems enough, the expansion of space would mean that, as the package traveled futureward, it would have to traverse ever-greater distances.

A mass shell will subject whatever it holds to time dilation, but it cannot hold much. The most massive mass shell possible would enclose a space a few meters across, thus posing a rather severe upward limit on the amount of energy or matter that might be preserved. Of course, a suitably advanced civilization might simply use more shells. But there would be other difficulties. Dismantling the shell to retrieve the package upon arrival (like slowing a package moving at relativistic velocities) would require a great deal of energy at a time when it is scarce. The scheme faces a more distant but rather more severe limit. By the time the universe is 10^{14} years of age, all matter will have degenerated. Because the shell, too, is made of matter, it cannot survive, and so cannot preserve its contents, longer than this.

However, certain stellar phenomena would outlast the decay of ordinary matter and might also serve the purpose. These are, of course, supermassive black holes, expected to be the most long-lived of all stellar objects. A black hole of a billion solar masses might survive to a time when the universe is 10^{100} years old. How might it be used for long-term storage? Recall that something falling toward the event horizon would experience time dilation. If the "something" is living, the fall will ultimately prove fatal (unless, of course, we countenance Frolov and Novikov's "wormhole rescue" from Chapter 5), and if the something is inanimate it would be irretrievable—and at any rate probably ripped apart.

Recall, though, that a person or package sent into orbit *around* a black hole is also subject to its gravitational field, and thus would likewise experience time dilation. Moreover, the degree of dilation could be controlled because it would be determined by the diameter of the orbit: the smaller the diameter, the greater the dilation. Over the very long period of time that we are contemplating, the black hole would lose mass, and the package would have to adjust its orbit accordingly. This challenge might be overcome. A more difficult problem would be presented to the civilization accepting delivery. Like dismantling the mass shell, retrieving the package from an orbit around a black hole would require a great deal of energy.

Krauss and Starkman suggested that, if wormholes exist naturally, a suitably advanced civilization might exploit them to transport materials and energy across otherwise vast distances. The authors did not suggest what might be a next step—that is, time-dilating one mouth so as to create a closed timelike curve. If a suitably advanced civilization desired a kind of "subjective" immortality —that is, surviving in a frame of reference that was increasingly separate from the outside universe—they might send themselves (or, perhaps more realistically, their representatives) through the wormhole to emerge from the "past" mouth, let them live and develop outside the wormhole until the moment they could reenter the "future" mouth and emerge from the "past" mouth again, ad infinitum.

Martin Rees notes that, because the universe is limited in complexity (the upper boundary set by the cosmic horizon, the lower boundary set by the inherent "graininess" of space), an intellectually curious civilization would be faced with an absolute limit to what it could learn. "The best hope of staving off boredom," Rees says, "would be to construct a time machine and, subjectively at least, exhaust all potentialities by repeatedly traversing a closed timelike loop."[14] (We might imagine a civilization aware of its own history, directing its own improvements on each circuit—a large-scale version of Harold Ramis's 1993 film *Groundhog Day*, in which a television weatherman relives the same day over and over and finds himself moving toward a personal enlightenment.) But Rees does not elaborate, and it is difficult to know how serious he is.

Perhaps there is another way a suitably advanced civilization might use a time machine to forestall its end. Suppose a wormhole time machine could be made to bridge spacetime from an era when resources of matter and energy are plentiful to a time when they are scarce. Might it then be used to deliver those resources? Perhaps, but in creating the time machine the civilization would need to subject one mouth to time dilation. As with the package, the energy required must be provided by the outside universe, and the more the mouth is time-dilated and the farther futureward it travels, the less energy the outside universe has available. In principle, one could time-dilate the mouth by other means—placing it inside a mass shell or in orbit around a black hole—but (again as with the package) retrieving the mouth from either would require a great deal of energy at a time when, by definition, energy is scarce.

Finally, all such schemes are thwarted by an energy deficit in the universe's future. But what if the wormhole were used to supply its own energy? Rather than bridging the span with a single wormhole, suppose there were a series—a Roman ring say, made into a "Roman string"? With judicious planning, the suitably advanced civilization might create a wormhole time machine and wait outside its time-dilated mouth for resources from the past sent (through the non-time-dilated mouth) by its pastward counterparts. Those resources could then be used to create a second wormhole, and to time-dilate a second mouth, and so on—the completed string operating like a series of pumping stations sending water uphill.

By such means, one imagines, a civilization might extend its lifetime appreciably. Because available energy and matter will diminish over time, if the civilization uses the same amount of energy for each wormhole, then it may expect the throat of each to be shorter than the previous. If the civilization wishes to make throats of the same length, it will require more energy with each construction. In either case, though, the string cannot be extended forever, because at some point resources will be exhausted completely.

Finally and inevitably, the universe will become uninhabitable. If the sufficiently advanced civilization is to survive, it must escape the universe altogether. The scientists introduced in previous chapters have demonstrated considerable imagination. It should come as no surprise that one of them—Edward Harrison, the astronomer who in Chapter 9 likened the universe to an oil painting—has described in general terms how such an escape might be accomplished. More provocatively, he has also suggested that it may already have *been* accomplished.

The Great Escape

Recall the "universe-in-a-lab" of Guth and Farhi. We viewed the experiment from the parent universe and left off shortly after the creation of the *child universe*. Let us shift perspective. The word "meanwhile" is a term whose meaning becomes deeply problematic here, but it will have to do. So—*meanwhile*, the child universe would begin to evolve as did our own—that is, inflating at a rapid, in fact exponential, rate. Or so we would expect, because we could never be certain. At the moment that the connecting wormhole is severed, the child universe is utterly separated from the mother universe, forever beyond the possibility of direct observation. Guth and Farhi could nonetheless make statements as to its nature. They predicted that the child universe would likely inherit the physical constants of its mother universe, with at most small changes.

What Guth and Farhi meant by "physical constants" are evident absolutes of nature that are fundamental (or seem to be)—so much so that we nonphysicists seldom think of them. A short list would include the speed of light, the strength of gravity, the masses of atomic particles, and the electrical charges and strengths of various interactions—that is, strong, electromagnetic, weak, and gravitational. All have specific values, and as far as we know they are the same everywhere in our universe, and they have not changed since its beginning. Yet they are a puzzle. Physicists know of no reason why the constants should have the values they have, and physicists have no theory that explains them, individually or collectively. But the situation is more interesting—or disturbing—still.

If one imagines the universe starting over with even one of the values adjusted up or down a fraction, life could not have appeared. Had the strong interaction between neutrons been slightly weaker, then the heaviest atom possible would be hydrogen, and we would have a universe of hydrogen, a place in which chemistry, and hence biology, would be impossible. Had the force of gravity been a fraction stronger, stars would collapse into themselves; a fraction weaker, and they would be unable to form to begin with. Had the electromagnetic force been slightly weaker, atomic nuclei would be unable to form. In fact, if even one of the many other constants were adjusted up or down even slightly, the universe as we know it simply could not exist. It is possible that a yet undiscovered natural law sets or controls the constants, but to appeal to this law for an answer would be begging the question, because we would have to ask what determines that law. Our existence, it would seem, is either a miracle or a fluke.[15]

The exactness of the constants has created intellectual crises of sorts for many scientists, including Fred Hoyle. In the late 1970s, Hoyle was researching the resonance states of carbon atoms. Carbon is the fourth most abundant element in the universe and the basis of all terrestrial life, yet it seems to have come into existence against all odds: were the carbon resonance level only 4 percent lower, carbon atoms would not form. Hoyle said that his atheism was "shaken" by the discovery. In 1981 he told an audience at Caltech, "A commonsense interpretation of the facts suggests that a superintellect has monkeyed with physics, as well as with chemistry and biology, and that there are no blind forces worth speaking about in nature."[16]

The Anthropic Principle

As it happened, it was Hermann Bondi and Thomas Gold—Hoyle's colleagues and coauthors of the *steady-state theory*—who had unintentionally inspired an idea that might allow Hoyle to hold on to his atheism. The *cosmological principle*, which is well established, states that on a large scale the universe looks the same in all directions for an observer at any place—or, more formally, that the appearance of the universe is independent of the observer's position in space. In 1968, Bondi and Gold broadened the principle to what they termed the "perfect cosmological principle"—that the appearance of the universe is also independent of the observer's position in time.

By implication, the principle provided support for the steady-state theory: if we assume that the universe always looked much as it does now, we must also assume that there can have been no big bang. Physicist Brandon Carter (the same person who in the mid 1960s had suggested that the interiors of black holes contain closed timelike curves) thought this was taking things a bit too far. He admitted that, although it was true that the conditions at the moment of the big bang made our presence then impossible, it did not follow from that fact that there was no big bang. He responded with the *anthropic principle*, which stated that "what we can expect to observe must be restricted by the conditions necessary for our presence as observers."[17]

There are two versions of the anthropic principle: weak and strong. The *weak* version states that the conditions necessary for the development of intelligent life will be met only in certain places or times. Intelligent beings at those places or times should not be surprised if they observe that their locality in the universe satisfies the conditions necessary for their existence. The *strong* anthropic principle states that the universe itself must have those properties that allow life to develop within it at some stage in its history.

Critics of the strong version of the anthropic principle suggest that it relies upon a suspect post hoc, ad hoc logic; they point out that we are intelligent life, and so it is inevitable that we find ourselves in a universe conducive to its (that is, our) existence. Nonetheless, if we win the lottery we may be justified in wondering why. Or, in a darker analogy, if we face a firing squad and every gun misfires, leaving us unharmed, we may be forgiven for taking a moment to ponder the reason.[18] In fact, the odds against all the constants having values conducive to life are many magnitudes greater than the odds against fifty guns misfiring simultaneously. Penn State physicist Lee Smolin estimates that the chances of randomly chosen constants in elementary particles and cosmology leading to a carbon chemistry necessary for life are one in 10^{220}.[19]

There is a third explanation for the values of the constants, one that followed from Guth and Farhi's prediction that a child universe would inherit the constants of its mother universe. In 1995, cosmologist Edward Harrison, then at the University of Massachusetts, speculated that our universe is artificial, created not by a god—or at least what we usually mean by a god—but by an intelligence superior to ours and existing in a universe whose physical constants are similar to our own.[20] Harrison's idea recalled a hypothesis made years earlier by Smolin. Smolin had suggested that black holes spawn new universes, and that universes with black holes come to outnumber universes without them. On a similar premise, Harrison constructed what he called a theory of "natural selection" of universes, but the reproductive mechanism of choice was not black holes. It was suitably advanced civilizations.

The "Natural Selection" of Universes

Harrison posited that universes that are unfit for life cannot produce the sufficiently advanced civilizations necessary to spawn child universes. Universes that are fit for life can produce sufficiently advanced civilizations, and these may, in turn, produce child universes. It was an argument with considerable explanatory power, and it satisfied the perfect cosmological principle. If universes whose constants do not support life are not reproduced, there should be relatively few of them, and they should represent a significant proportion of the total number of universes only when the multiverse is young. If universes whose constants do support life multiply and multiply again, there should be a great number of them, and as the multiverse grows older that number should increase exponentially, with fertile universes coming to greatly outnumber sterile universes.

So, if most universes are conducive to life, it follows that we are not living in a privileged place, but just another universe—in fact, rather a typical one. Furthermore, if at most times in the history of the multiverse there are a great number of universes, then we are not living at a privileged time either. In fact, since there are more universes with each generation, odds are far greater that the universe we inhabit is a member of a late generation than a member of an early generation. Harrison had taken the problem presented by the physical constants and turned it upside down. Suddenly the chances were greatly in favor of our finding ourselves in a universe with constants fine-tuned for life.[21]

Most of the universe is so distant that we cannot see it; the universe has lasted for billions of years and will last for many hundreds of billions more. We might reasonably expect to find it so alien as to be forever beyond our ken. But it is not. Einstein said, "The eternal mystery of the world is its comprehensibility."[22] Indeed, the inventions of Euclid, Descartes, and Riemann can be used to explain the universe, and to explain the shape of space itself. Exactly how we have managed to understand so much is a bit of a mystery, and Harrison suggests an answer. If the universe had designers, then its designers thought as we do. If its designers thought as we do, it should not surprise us that we can comprehend their work.

Might we pursue this line of thought further—and suppose that we might also understand their motives? Harrison did. He went so far as to speculate on the reasons that a civilization might undertake such a project, and he imagined three: first, an experiment for its own sake, simply to see whether it can be done; second, after the fashion of Stapledon's Eighteenth Men, to create an abode for life and intelligent life, even if it is a place they can never see, let alone visit; third, to create a place that they might themselves inhabit—this assuming, of course, that the civilization will have found a means to hold an intra-universe wormhole open and traverse it.

As to how all these reproducing universes began, Harrison suggested two possibilities. Perhaps there was a single first universe whose physical constants were conducive to life, and it produced a civilization that produced fertile universes that, in turn, produced other fertile universes. Or perhaps there was an initial ensemble of universes with a wide range of physical constants. The universe whose physical constants were conducive to life produced a civilization that created another universe, its physical constants also conducive to life. Fertile universes came to outnumber sterile universes, and the proportion increased over time.[23]

A MESSAGE FROM ANOTHER UNIVERSE

J. Richard Gott and Li-Xin Li speculated that a civilization creating a child universe might be able to adjust its physical constants so that they contained a message. A message from the universe's architect and addressed to us is a rather arresting idea, and it should come as no surprise that figures mentioned earlier in this book have given it serious thought. Andrei Linde spoke to author Timothy Ferris about how a written message, tossed through the wormhole umbilical somehow held open would, in the child universe, quickly grow to so great a size that it would stretch far outside the cosmic horizon of anyone who might try to read it.[a]

Sagan's novel *Contact* concludes with a suggestion that a sort of signature is encoded deep in the infinite string of (presumably) random numbers in pi, the ratio of a circle's circumference to its diameter. (The logical difficulty with Sagan's concept is that pi is not a physical constant, and not a property of this particular universe; changing the value of pi would be like changing the value of 2.) But let us suppose it is possible. Why should the supercivilization leave a signature to begin with?

An artist's signature is part appeal to vanity, part appeal to commerce, alerting its readers to the identity of a maker about whom they can presumably learn more. But if, as Gott and Li suggest, it is *just* a signature and not, say, a thousand-volume history of their civilization, then it could serve neither of those ends. The reader of the signature could never learn more about the civilization; it existed, after all, in a universe forever inaccessible from her own. So perhaps the signer's motive would be more purely altruistic. If nothing else, a signature would prove that a civilization had once created a universe, and so would imply that it might be done again. In this way a universe-creating civilization might alert the civilization in the child universe to its own potential, and thus add some insurance to the continuance of life, and of universes.

[a] Ferris, *Whole Shebang*, 262.

Harrison acknowledged that both models leave a rather large question unanswered. If the first is true, then what created the first universe? If the second is true, what created the initial set of universes? Neither model addressed what is called the "problem of first cause." They merely pushed it further back in time.

First Cause

The questions surrounding a first cause disturbed Aristotle so much that he preferred to imagine a universe that existed eternally; Einstein's first application of general relativity to cosmology produced a model of a static universe with neither beginning nor end; and proponents of the steady-state theory counted among its virtues that it required no first cause. In the early part of the twentieth century, the problem, to the relief of many, was avoided. But the rise of the big bang cosmology, which suggested that the universe had a finite age, put the problem front and center. Then, in the 1960s and 1970s, Penrose and Hawking demonstrated that, if Einstein's field equations are correct and if we take the current measured expansion and extrapolate backward, we arrive at an initial singularity. Some began to speak of *it* as the first cause. Naturally, questions arose as to what was going on *before* the singularity, and the standard answer, as old as Plato and Augustine, was that the question has no meaning because time was created at the moment space was created. There was no "before."

Even here there remained a problem. The cosmic microwave background is radiation left from the big bang, now so greatly redshifted that it appears not as light but as microwaves. The background is a "fossil" of the initial singularity, and although it is very nearly uniform, it exhibits very small fluctuations—one part in a hundred thousand. The universe would not be as it is (indeed, galaxies and large-scale structures could not have formed) if the initial singularity had had a uniformity other than this—that is, very nearly perfect, but slightly less than perfect. We might think of this uniformity as the first physical constant, and it brings with it the attendant problem: it is quite precise, and yet there is no particular reason it should be what it is, except that (and here we are flirting with circular reasoning) if it were otherwise we would not be here to wonder about it. A tuning mechanism would have to precede the creation of the singularity. It seemed we needed a "before" after all.

In the well-known and possibly apocryphal story, a woman informs a philosopher (it is Bertrand Russell in some versions, psychologist William James in others) that the universe rests on the back of a turtle, and that the turtle rests on a still larger turtle. The philosopher responds by asking her what *that* turtle rests on, perhaps hoping that she will see the flaw in her reasoning. But she answers, "It's no use, Sonny. It's turtles all the way down."

There is a scientific version of the turtle model, termed the *oscillating universe*. It depicts universes strung out through time in an infinite series. Each in turn expands and contracts into a big crunch, then rebounds into a new universe. Although now mostly discredited, the model had its virtues. For one, it explained the less-than-perfect uniformity of our universe's initial singularity by proposing that it was set in a previous universe that began with another singularity, also with a less-than-perfect uniformity, and so on. However, it did not address the problem of first cause any more than the turtles did. Moreover, it proved to have a number of internal problems—one of which was that universes must begin in states of low entropy and grow more disordered, and no one could explain how the great disorder during the crunches could be recycled into the very low disorder of the subsequent bangs.

In 1982, Stephen Hawking and cosmologist James Hartle proposed a model that required no moment of creation. Rather, near the universe's beginning, space and time together compose a four-dimensional surface with no boundary and no edge. Hawking declared, "The universe would be completely self-contained and not affected by anything outside itself. It would be neither created nor destroyed. It would just BE."[24] Then, in 1994, Alexander Vilenkin of Tufts University suggested that the universe might have come into existence through quantum tunneling, a process by which a quantum system may suddenly transition from one state to another.

Gott and Li Consider First Cause

In the early 1990s, J. Richard Gott had remained interested in the possibility of closed timelike curves, and the fun of thinking about time travel was not lost on him. One Christmas he received a sport jacket as a present. Gott has described its color as that of a deep lake; others have called it an odd sort of opalescent sea green. He first wore it at a conference in California, expecting that it would blend in with native dress. He was wrong. Harvard astronomer Bob Kirshner joked that the color didn't exist in the present and could only have come from the future. They had a laugh over it, but Gott began to wear what he called the "coat from the future" whenever he lectured on time machines, and he lectured often.

It was Gott's ongoing interest in time travel that provoked a letter from Li-Xin Li, the undergraduate who in 1994, recall, suggested that a reflecting sphere between wormhole mouths would preserve the closed timelike curve. Gott had admired the piece, and with Gott's recommendations, in 1997 Li was admitted to Princeton as a doctoral candidate.

Together, Gott and Li began to think about the problem of first cause. They noted that none of the proposed beginnings described in the previous section were truly self-contained. Hawking and Hartle's model began not with nothing, but with a quantum state. Likewise, the place from which Vilenkin's universe tunneled may have been a state with no classical spacetime, but it was nonetheless a state. If the universe had been created by a quantum fluctuation (after Tryon's suggestion in 1973), then one might reasonably ask, a quantum fluctuation of *what?*[25]

Genesis as Time Machine

Still, there was one origin theory, Gott realized, that might resolve the problem rather nicely. It was Andrei Linde's "chaotic inflation," in which inflating universes give birth to other inflating universes, like branches growing from a single trunk. There were as many bangs as branches, and necessarily, there was a first bang that produced the trunk. As to what might have preceded *that*, Linde had no answer. But Gott and Li suggested one. They observed that during the Planck epoch, the universe was made of quantum foam, and quantum foam includes all topologies, among them closed timelike curves. Therefore, it was conceivable (and in fact, sooner or later inevitable) that a branching universe circled around to become the trunk.

The primordial loop that Gott and Li proposed would contain "a pure inflationary vacuum state with a positive energy density and negative pressure everywhere,"[26] and this state was achievable if the loop had a certain length—a very short one, measured in fractions of a nanosecond. Gott and Li also suggested the possibility of a second loop, occurring somewhat later, during the epoch of grand unification. In either case, they claimed, at the very beginning of the universe we might expect to find no first cause per se—but a closed timelike curve. The universe did not exist eternally, and neither was it created out of nothing. Through a closed timelike curve, it gave birth to itself.

Gott and Li were pleased with the solution and admitted surprise upon discovering that their model answered *another* question—one they had not even asked. Recall the mysteries associated with time's arrow. Electromagnetism, for instance, is indifferent to direction in time; this means that, in principle, light from the present moment could travel all the way back to the big bang.

Yet we never see light traveling to the past, even in shorter trips. Physicists cannot explain why.

Gott and Li calculated that, during its circuit away from and back to the trunk, the loop would increase in circumference more than five hundred–fold; naturally, the spacetime within it would be stretched as well. Recall how schemes for sending resources futureward were made difficult because stretched space drains energy from whatever is traveling through it. If light, imagined here as waves or particles, traveled through the loop toward the future in, say, the clockwise direction, its energy would be diminished on each circuit. However, light that is traveling pastward—in the counterclockwise direction— would have its energy *increased* on each circuit, and soon there would be an infinite buildup of energy that would destroy the loop.

It seems, then, that a closed timelike loop at the beginning of the universe would not allow energy to travel pastward. So although the laws of physics do not forbid light (or any electromagnetic waves or gravitational waves) from traveling pastward, the geometry of space- time at the universe's beginning does. In Gott and Li's solution, it was that geometry that set the direction of time's arrow.

At least in outline it was an elegant solution, and it certainly made for an elegant embedding diagram. It looked like trumpet bells sprouting from a single barrel, and the barrel looping into itself. In his book *Time Travel in Einstein's Universe*, Gott mentions that his wife remarked that the diagram recalled a fanciful con- struction of Dr. Seuss. Gott himself saw in it a strange beauty, enough that he sought out a New Jersey glassblower and asked him to produce a small handheld version. Now whenever Gott speaks on time travel, he wears a coat from the future and shows a model of the past.

By the standards of *Physical Review D*, Gott and Li's paper was long, with room afforded the authors to devote a full page to specu- lation not usually seen in that journal. Gott and Li noted that the time loop may be natural; or it might have been produced artificially, in which case "all of the individual universes would owe their birth to some intelligent civilization,"[27] which might be us—or our descendants.[28]

Gott has acknowledged that his model of genesis via time machine may be interpreted as encroaching upon the territory of philosophers and theologians. (Gott is a practicing Presbyterian but observes a separation between his work and his religion.) However, he also notes that the model is not so audacious as it might seem. It does not even try to answer a deeper question much considered by philosophers—namely, "Why is there something instead of nothing?" or "Why is there a universe at all?" Finally, then, the model allows for something "outside" the multiverse. It leaves room, one might say, for Peter Damian's god.

EPILOGUE

"I know," he said, after a pause, "that all this will be absolutely incredible to you, but to me the one incredible thing is that I am here to-night in this old familiar room, looking into your friendly faces, and telling you all these strange adventures."
— H. G. Wells, *The Time Machine*

Physicists first investigated pastward time travel nearly eighty years ago, and since the late 1980s they have made it the subject of more or less sustained study. In that time they have learned a great deal. They have found (and been surprised to find) that general relativity allows for pastward time travel through an effect called frame dragging. They have learned that all schemes that would employ this effect are unphysical—that is, impossible in the universe we know—but that several other designs, although far beyond the reach of current technology, may be within the realm of the possible.

They have learned that a straightforward means to create a time machine—one possible in our universe—is to produce a wormhole and subject one of its mouths to the time dilation effect described by special relativity. They have learned that the wormhole itself is a physical construct allowed, at least in principle, by Einstein's field equations, and that its conversion into a time machine might occur inevitably and spontaneously. They have learned that cosmic strings might also be used to create a time machine, although this design, like those that would use frame dragging, may well be unphysical.

Their inquiry has reached into areas outside the traditional domain of physicists and cosmologists, encouraging more thorough explorations of the philosophical issues of causality and free will. They have found that time travel need not introduce causal paradoxes, and in fact might be circumscribed by a principle of self-consistency that is in itself explained by the sum-over-histories formulation of quantum mechanics. They have learned that spacetime might be warped in ways that would (again, in principle) allow the creation of a variety of time machines. They have discovered that negative energy would be a necessary component of all of these, as well as of any imaginable time machine existing in a compact space, and they have also found that negative energy might prove impossible to produce in the quantities necessary. They have been reminded that most of their schemes assumed a traditional model of spacetime, and that spacetime might be structured in such a way that pastward time travel, if possible, would occur "across" universes, although the assumptions underpinning this scenario have been seriously questioned. Finally, and perhaps most significantly, they have raised questions that are helping to discern the outlines of a theory of quantum gravity—the "theory of everything" that promises the unification of all physical laws.

The inquiry into closed timelike curves, far from being a steady progression, has seen advances and retreats. And in this sense it is rather like science itself. It is easy to imagine our understanding of the universe as a patch of light on an otherwise darkened stage. Over time the patch of light grows, then shrinks a bit, then grows again, perhaps a bit larger. At some future moment, we imagine, ignorance will be banished entirely, and the whole stage will be lit. Perhaps. But in 1444 Nicholas of Cusa suggested that this conception was overly simple and put forth what he thought to be a more accurate model of learning—one that recognized two types of ignorance. The first he called "unlearned ignorance." By this he meant a simple and straightforward lack of knowledge. The second, what he called "learned ignorance," is a somewhat subtler quantity. It is *what we know we do not know*. Learned ignorance, so Cusa argued, is inextricable from knowledge, and in fact grows alongside it. The more we learn, the more we learn how much we do not know. It may be that the most significant finding of the inquiry into closed timelike curves are two coupled pieces of learned ignorance: the facts that we do not know whether nature protects causality, and that we will not know until we develop a full theory of quantum gravity.

Since the late 1990s, work on closed timelike curves has fallen off quite a bit. Many of the physicists who drove the inquiry, sensing the trail gone cold, have turned their attentions elsewhere. This is not to say it has been abandoned entirely. In 2003, Visser and two colleagues found that it is possible to construct a wormhole with only small amounts of negative energy;[1] in 2004, Thomas Roman published a piece on negative energy and wormholes, reviewing recent progress in those areas;[2] and in 2005, Amos Ori suggested that causality violation might occur in an "empty torus."[3]

Physicists have a tradition of celebrating sixtieth birthday parties, and both Thorne's and Hawking's have come and gone. They were well-attended affairs. Many figures mentioned in this story were present, and as might be expected, there were jokes about getting older and somehow escaping from aging with time machines.[4] Meanwhile, a next generation of physicists has arisen, and they have approached the question from new directions. A research group at Queen Mary, University of London, used a toy model of string theory to find solutions that would allow closed timelike curves. Petr Horava at Berkeley tasked his graduate students with discovering ways in which more sophisticated models of string theory might undermine those results. They succeeded, identifying chronology protection mechanisms in several spacetimes, including Gödel's rotating universe. In 2004, Lisa Dyson, then a graduate student at MIT, considered causality violation outside the event

horizon of a rapidly spinning theoretical cousin to the more familiar black holes.[5] She concluded that it could not spin at a rate fast
enough to generate closed timelike curves, although it could
approach that rate very closely. Her result recalled the manner in
which Hawking's proposed protection mechanisms stop the creation of closed timelike curves at the last possible moment. Indeed,
one of Dyson's colleagues observed, "It's as though you're trying to
build that last little bit of your time machine and there's a force that
stays your hand."[6]

So far, the results suggested by string theory, like those given by
quantum mechanics and its derivations, are that various types of
time machines are prevented by various mechanisms. Because
most physicists are predisposed to economy and elegance, this multiplicity of results is unsatisfying. They would prefer a single mechanism that works for all cases.

I began this book with a promise to distinguish science from science
fiction, and a caveat that a significant sort of creative cross-
pollination was at work. Some of it, as we have seen, has been
direct. Carl Sagan and Robert Forward, seeking to ground their
own science fiction projects in actual physics, motivated the inquiry
into closed timelike curves at crucial moments. But much of the
influence has been indirect. The earliest incarnations of Novikov's
self-consistency principle, as well as its most colorful illustrations,
are from science fiction. Certainly the issues that arise in thinking
about closed timelike curves—the nature of a sufficiently advanced
civilization, time travel paradoxes, and "parallel" universes, to
name a few—call upon a sort of imaginative play.

In fact, many of the physicists involved in the inquiry enjoy science fiction. Thorne told an interviewer in 2000 that he reads a science fiction novel during his vacation every year. Roman teaches a
course on the nature of time that has several time travel short sto-

ries on its syllabus. Although the photographer for the picture of Novikov shown in the insert told me he supplied the copy of Hergé's *Destination Moon*, clearly Novikov is enjoying it. Because I was curious about the precise role of science fiction in this story, I was especially intrigued when a few of the physicists I interviewed mentioned that, at Aspen in 1992, they had indulged in conversations concerning time travel that ranged widely and were unrestrained by scientific method. To the regret of some and (no doubt) relief of others, those interchanges were unrecorded. But at least one of the wilder ideas tossed about during those two weeks *did* appear in print. It was Lossev and Novikov's clever spacecraft.

The idea is seldom cited in the professional literature, and so I was curious as to its reception at Aspen. Matt Visser told me that it was received "lightheartedly," yet went on to note that the idea upon which it is based—the self-consistency principle—is in some ways the backbone of the 1990 "consortium" paper.[7] That work, coauthored by seven physicists (one of whom was Novikov himself), is regarded as one of the more important papers on closed timelike curves. It would seem—and this should come as no surprise—that the subjects of this book are able to hold the intellectual playfulness typical of science fiction in a sort of critical suspension, and temper and channel it by the more rigorous sort of thought that we call the scientific method.

Although many of the people described in this book read science fiction in their youth, they had a thirst for a deeper knowledge. At some point there was a shift in expectations, a kind of intellectual maturation. One figure likened that moment to seeing a landscape painting removed from a wall to reveal, behind it, an open window. At first the view out the window was not as spectacular as the scene in the painting; then he came to understand that it only *seemed* less spectacular. He needed to learn how to see it. It is much the same with the others. Listen to any of them speaking, and you will realize that they are not so excited by imagined universes as they are by real universes—by what *is*. If it happens that quantum gravity allows time machines, they will regard the universe as a strange and wondrous place. If it happens that quantum gravity forbids time machines, they will regard it as no less strange and wondrous.

And they will begin to ask new questions.

NOTES

Prologue

1. For an overview of the social and intellectual context from which Wells's ideas were drawn, see Leon Stover's introduction to the 1996 edition of Wells's book.

2. It should be said that few, if any, of the scientists discussed here would regard time machines as a focus of their ongoing research. Thorne's interests have been black holes and, more recently, the detection of gravitational waves; Hawking's primary interest is quantum gravity and the origin of the universe; much of Tipler's work has focused on gravitational physics and general relativity. The one exception may be Matt Visser, now at Victoria University of Wellington, New Zealand. He has written more extensively on time machines than anyone else, and many of his other interests—among them the Casimir effect and quantum aspects of black holes—bear closely on that subject. But Visser, too, has done work in areas less directly related to the subject—for instance, the cosmological constant. All of these scientists were at some point drawn to the subject of time travel, either as a consequence of their own studies or as a ready means to gain a handle on a separate problem.

3. Einstein, *Expanded Quotable Einstein*, 283–284.

4. Wells, *Outline of History*, 758.

5. Readers seeking histories of time travel in science fiction are advised to consult Paul Nahin's *Time Machines: Time Travel in Physics, Metaphysics, and Science Fiction*; and relevant sections in Marilyn P. Fletcher, ed., James

L. Thorson, consulting ed., *Reader's Guide to Twentieth-Century Science Fiction* (Chicago: American Library Association, 1989); as well as John Clute and Peter Nichols, *The Encyclopedia of Science Fiction* (New York: St. Martin's Press, 1993).

6. Quoted in Harrison, *Cosmology*, 162.

Chapter One: Intimations of Spacetime

1. By many accounts, the first fiction describing pastward time travel was the 1891 *Tourmalin's Time Cheques*, a charming work by "F. Anstey" (a pseudonym for Thomas Anstey Guthrie), a novelist known for his humorous critiques of the British middle class. The story begins when the weak-willed and dull-witted Tourmalin of the title, who wishes that his day would pass faster, is met by a representative of the "Anglo-Australian Joint Stock Time Bank, Limited," a firm that allows clients to deposit time in amounts of durations no shorter than fifteen minutes, and to withdraw it, with added interest, at their convenience. Tourmalin, with a book of checks, is whisked months into his future, and soon he begins to cash his checks. Perhaps because they were printed out of sequence, he is transported to various times in his past, also out of sequence, and so his experiences are quite bewildering. He discovers that, despite his recent engagement to marry, he is romantically involved with two women. After barely escaping assault from a rival suitor and the wrath of his betrothed, Tourmalin awakens, relieved to discover that he dreamed the whole affair.

2. Carroll, *Complete Sylvie and Bruno*, 150.

3. In 1877, London society was fascinated with a spiritualist named Henry Slade, who claimed to be in contact with spirits from the fourth dimension. Slade was put on trial for fraud, and many respected scientists defended him. One particularly gullible supporter was German astronomer and physicist Johann Friedrich Zöllner. Zöllner offered Slade the opportunity to prove his claims; indeed he hoped Slade would be able to do so. He challenged Slade to interlock two unbroken rings made of different woods without breaking them, and to turn a right-handed spiral snail shell into a left-handed one. We may appreciate how such operations were supposed to offer evidence of a fourth dimension if we imagine corresponding actions in a two-dimensional universe—that is, a plane. Cut a horizontal cross section from each ring and place the slices side by side on a flat surface. They may slide around each other, they may bump against each other, and they may come to rest with their edges touching—but as long as they remain entirely within the plane, neither can be superimposed on the other. The only way we can overlap them is by first lifting one slice out of the plane. The snail shell is a similar case. Cut a cross-section of a right-handed shell and place it in the plane. No amount of sliding the cross section around or rotating it within the plane will ever make it left-

handed. The only way it can be translated into its left-handed counterpart is if it is lifted from the plane, turned over, and dropped back into place. Although Slade was eager to perform many feats that he had not been asked to perform, he could not interlock the wooden rings and could not translate the shell. Nevertheless, Zöllner's faith in Slade, perhaps sadly, was unshaken.

4. Newcomb authored several pieces on the subject, arguing (contra Hinton) that we in the third dimension cannot properly imagine a space of two dimensions, let alone four. He wrote, "As illustrating the limitation of our faculties in this direction, it is remarkable that we are unable to conceive of a space of two dimensions otherwise than as contained in one of three. A mere plane, with nothing on each side of it, is to us inconceivable." ("Modern Mathematical Thought," 325.)

5. Wells, *Experiment in Autobiography*, 172.

6. C. H. Hinton's "What Is the Fourth Dimension?" had been reprinted in the *Dublin University Magazine*, the *Cheltenham Ladies' College Magazine*, and in 1884 in a pamphlet with the rather provocative subtitle "Ghosts Explained."

7. Grosskurth, *Havelock Ellis*, 102.

8. Gauss, *General Investigations*.

9. *Flatland* is presented as the autobiography of "A. Square," a two-dimensional being who inhabits a two-dimensional universe that he calls, for the convenience of his three-dimensional readers, "Flatland." (Some have suggested that "A. Square" may be a play on his creator's unusual full name—"Edwin Abbott Abbott," or "Abbott squared.") The work divides itself into two sections. The second section is social satire. The first section, parts of which are excerpted in many geometry textbooks, is an exposition of ideas of the nature of space. It describes A. Square's education regarding worlds of various dimensions. His teacher is "the Sphere," who enters Flatland as a sphere might enter a plane—from above it. To A. Square of course, the Sphere appears magically, as if from nowhere. As A. Square watches the Sphere enter that plane, at any given moment he can see only a slice that he perceives as the edge of a circle. The circle grows as the Sphere descends into A. Square's line of sight and shrinks as it rises out of it. A. Square is shown how such movement is accomplished, his education greatly assisted through analogies with dimensions below his own—that is, Pointland and Lineland. Abbott's readers in the third dimension are thereby invited to draw corresponding analogies to dimensions above their own. Abbott, *Annotated Flatland*.

10. *Flatland* has seen numerous imaginative reworkings. Dionys Burger's *Sphereland* (written in Dutch in 1957 and translated into English in 1965)

is narrated by A. Square's grandson, A. Hexagon. A. K. Dewdney's *The Planiverse* (1984) describes the physics and biology of a two-dimensional universe. Jeffrey R. Weeks's *The Shape of Space* (1985) weaves tales of Flatland into a text on topology. Rudy Rucker's *The Fourth Dimension* (1985) extends Flatland in several directions. In Ian Stewart's *Flatterland* (2001), A. Square's great-great-granddaughter, Victoria Line, explores higher-dimensional geometry and quantum physics.

11. Found in Hinton, *Speculations*.

12. Nearly two decades later, in 1907, Hinton extended these ideas in "An Episode of Flatland: Or, How a Plane Folk Discovered the Third Dimension" (in Hinton, *Speculations*).

13. The first use of the term "tesseract" seems to have been by Hinton, in his 1888 work, *A New Era of Thought*.

14. C. H. Hinton, "What Is the Fourth Dimension?"

15. It is interesting that a few years later Wells used a similar fantasy in a short story: "'It is confusing,' said the Vicar. 'It almost makes one think there may be (ahem) Four Dimensions after all. In which case, of course,' he went on hurriedly—for he loved geometrical speculations and took a certain pride in his knowledge of them—'there may be any number of three-dimensional universes packed side by side, all dimly dreaming of one another. There may be world upon world, universe upon universe.'" *Wonderful Visit*, 25–26.

16. Mathematicians, on the other hand, are perhaps more comfortable with abstractions, and spaces whose dimensions number more than ten or eleven are rather familiar territory. In fact, the discipline of functional analysis was developed from the need for a theory of infinite-dimensional spaces.

17. Hinton, *Scientific Romances*, 8–9.

18. Rucker, *Speculations*, 41.

19. Hinton, "Mechanical Pitcher"; and "Mechanical Baseball Pitcher" (*Scientific American*).

20. Augustine, *Confessions*, 264.

21. Ibid.

22. Ibid. Augustine observed that the present is in a continual state of self-negation, and that its existence, strangely enough, depends upon its soon-to-be *nonexistence*.

23. *Ovid's Metamorphoses*, bk. XV: 176–198, *Pythagoras's Teachings: The Eternal Flux*. Translated by A. S. Kline, http://etext.virginia.edu/latin/ovid/trans/Metamorph15.htm (accessed August 22, 2006).

24. Marcus Aurelius, *Meditations*, bk. 4, sec. 43.

25. A clock is of no help here. It may indicate the passage of time, but it does not actually measure the rate of that passage. What we mean when we call a clock accurate is only that it is regular: it is true to itself or to an external source like, for instance, the resonance frequency of a cesium atom. But that external source also merely indicates and does not measure the passage of time.

26. The concept probably arose from the appreciation of beginnings and endings, and of stages of a human life. As theologians note, the Judeo-Christian belief that God manifests in human form at a specific point in history had much to do with the widespread adoption of the linear model of time by whole societies. Beginning perhaps in the nineteenth century, an awareness of the quickening pace of industrial or technological progress underscored this idea, suggesting that human history was a progression. Another model of time on large scales—rather at odds with this—is *cyclical* time, which suggests that all that exists moves through cycles that extend both into the past and into the future, ad infinitum. This model is inspired by the circuits of heavenly bodies, the seasons, and predictable natural events like seasonal river flooding and animal migrations. In our daily lives, most of us operate with both models, blending them according to our need and inclination.

27. The tenseless theory is also termed the "static theory," because it requires no movement: ideas of time moving through things, or things moving through time, are unnecessary.

28. Rucker, *Speculations*, 124–125.

29. Ibid., 125.

30. Ibid., 80.

31. Similarly, many have observed that Wells anticipated Einstein's special relativity by some ten years. Even in the most generous sense, such a claim is unsupportable. But it is unfair to criticize Wells's novel by faulting its physics. In 1895, ideas of space and time were only beginning to crystallize, and Wells himself nowhere suggests that he intends us to regard the theory that his novel employs as anything more than a pleasant, enabling fiction. In fact, he later wrote that *The Time Machine* and stories like it "do not aim to project a serious possibility; they aim indeed only at the same amount of conviction as one gets in a good gripping dream." (H. G. Wells, *Seven Famous Novels* [New York: Knopf, 1934].) Nonetheless, to understand why it cannot work may be instructive. Although Wells's *The Time Machine* seems to have used Hinton's concept of time as a fourth dimension, it did not borrow Hinton's filamentary atom. On the time axis, Wells's machine overtakes itself and actually passes through itself. Such

an action would be impossible in Hinton's model, in which one cannot accelerate one's duration. As we will see in Chapter 2, it is also impossible in spacetime as Einstein explained it and as we understand it.

Chapter Two: Einstein's Radical Idea

1. Quoted in Ferris, *Mind's Sky*, 92.

2. Mechanical clocks of various types were used as early as the third century BC, but from the sixth century AD to the first millennium we have no record of mechanical timekeepers in Europe. On that continent clocks seem not to have reappeared until around 1000.

3. Newton, *Mathematical Principles*, 6. Galileo used straight line segments to denote intervals of time, and Isaac Barrow, chair of mathematics in Cambridge, formalized this practice, offering a discussion of time represented as a straight line, saying, "For time has length alone, is similar in all its parts and can be looked upon as constituted from a simple addition of successive instants or as from a continuous flow of one instant; either as a straight or a circular line." (*Geometrical Lectures*, 35.) Barrow was Newton's predecessor in the Laucasian chair, and greatly influenced Newton.

4. Locke, *Essay*, bk. 2, ch. 15, par. 11.

5. Pais, *"Subtle Is the Lord—"*, 37.

6. Ibid., 38.

7. Ibid., 48.

8. Ibid., 48.

9. Maxwell, *Scientific Papers*, 763. The theory of the aether had a strong hold on the scientific imagination. Eddington's 1920 *Space, Time and Gravitation*, a summary of Einstein's special and general relativity, suggested that physicists might wish to retain the idea of the aether in modified form.

10. In the seventeenth century, Galileo attempted to measure the velocity of light with signals between lanterns positioned a mile apart. He failed, concluding that the speed was too great to be measured by that means. In 1675 the Danish astronomer Olaus Rømer achieved a result very close to the actual speed by measuring periodic variations in times at which Jupiter's moons were eclipsed by the shadow of the planet.

11. Pais, *"Subtle Is the Lord—"*, 126.

12. Recently, John Rigden suggested that, because Planck was the journal's advisor on theoretical physics, it was his favorable opinion of Einstein's work that allowed it to see print in the first place. Rigden, *Einstein 1905*.

13. Pais, *"Subtle Is the Lord—"*, 44. For his part, Einstein the student had regarded Minkowski as an excellent teacher.

14. Many of these formulations and terms grew from a paper that Minkowski published in that year. Minkowski, *Gesammelte Abhandlungen*, 352.

15. Lorentz et al., *Principle of Relativity*, 75. Sadly, Minkowski would not live to see the Cologne Lecture appear in print, and he would not live to see his illustration of special relativity embraced by Einstein. In 1909 he died of appendicitis at the age of forty-five.

16. Light and all electromagnetic energy travel through a vacuum at 299,792.5 kilometers a second, with an uncertainty of less than one part in a million. The speed happens to be so close to 300,000 kilometers a second that many scientists have seriously proposed that it be made the basis of the metric system of measurement.

17. When physicists speak of something traveling between two points in spacetime, they use the word "signal." A signal may be a fragment of Morse code conveyed by a radio wave, the image of a star, or a rock that hits the side of my head. That the thrower of the rock may mean to convey no articulate message is irrelevant, since some information is conveyed, even if that message is as (painfully) obvious as "a rock was thrown." Likewise, if the rock simply falls from the sky, then information is nonetheless conveyed, even if it is the mere fact that a rock has fallen from the sky.

18. Even objects that are near us—say, the furniture in the same room—are in our past. It is, to be sure, a very recent past—a small fraction of a microsecond ago—and for all ordinary intents and purposes, we might be sharing the same moment with it. Such is the case for most anything in our Earthly environment. In daylight the most distant object we can see is likely to be the Sun, and its image is about eight minutes old. Only when we look to the night sky do we begin to experience spacetime on appreciable and (in fact) enormous scales. We see the bright star Sirius as it was eight years ago, and most of the stars in the Milky Way Galaxy as they were centuries ago. The Andromeda Galaxy, generally regarded as the farthest object visible with the unaided eye, appears to us as a hazy patch of light. According to the most recent estimates, we see it as it was nearly three million years ago.

19. Lorentz et al., 76.

20. Notice that if we sliced the upper light cone in Figure 2.5 horizontally, we would get a two-dimensional representation with two spatial dimensions and no time dimension represented. It would appear as a circle and represent a wave front of light. In three-dimensional space that wave front would appear as a sphere moving outward from its center at the speed of light.

21. Mathematicians cannot visualize higher-dimensional forms any more

than you or I can. But with coordinate geometry they can define the properties of such forms with great precision. An illustration may be in order here. A three-dimensional sphere has x, y, and z axes. A four-dimensional sphere—a "hypersphere"—has x, y, z, and t axes. We cannot visualize this hypersphere, but using coordinate geometry we can actually calculate its volume. The area of a circle is πr^2. The volume of a sphere is $\frac{4}{3}\pi r^3$. The volume (or hypervolume) of a four-dimensional sphere, therefore, is $\frac{1}{2}\pi^2 r^4$.

22. Suppose we want to measure the distance between two points in spacetime. We make the distance between the two points the hypotenuse, or longest side, of a right triangle. One horizontal side of the triangle would be the space intervals, and the vertical side would be the time interval. In Einstein's equations the parameters representing space are positive, but the parameter representing time appears with a minus sign. We might roughly approximate them with this equation: (spacetime interval)^2 = (space interval)^2 − (time interval)^2. In the spacetime version of the Pythagorean theorem, the length of the hypotenuse is equal not to the sum of the squares of the other two sides, but to the difference between them.

23. From Newton's cosmos arose new ideas that, although God had created the universe, he did not intervene moment to moment but instead had retreated to a cosmic back stage. If God's influence was indirect, it followed that there could be no divine right of kings. Thus did physics and astronomy affect the more mundane, political sphere, allowing John Locke to assert, "The natural liberty of man is to be free from any superior power on earth, and not to be under the will or legislative authority of man, but to have only the law of nature as his rule." (*Second Treatise*, 17.) In ensuing years, the same idea appeared in new and unexpected places— the philosophies of Rousseau and Jefferson, the precepts of the Declaration of Independence, and the French Revolution. All owed a profound debt to Newton.

24. Where Newton makes this admission is an especially beautiful and powerful passage, worth quoting at length: "Hitherto we have explained the phenomena of the heavens and of our sea by the power of gravity, but we have not yet assigned the cause of this power. This is certain, that it must proceed from a cause that penetrates to the very centres of the Sun and planets, without suffering the least diminution of its force; that operates . . . according to the quantity of the solid matter which they contain, and propagates its virtue on all sides to immense distances, decreasing always as the inverse square of the distances. . . . But hitherto I have not been able to discover the cause of these properties of gravity from phenomena, and I frame no hypotheses. Hypotheses have no place in experimental philosophy . . . to us it is enough that gravity does really exist . . . and abundantly

serves to account for all the motions of the celestial bodies, and of our sea." *Mathematical Principles*, 547.

25. We should note that Einstein's reverence for Newton was profound. Lord John Scott Haldane, who acted as Einstein's host upon his first visit to England, said that, on the first morning of his stay, Einstein left his house to visit Westminster Abbey, intending to gaze on Newton's tomb. Pais, *"Subtle Is the Lord—"*, 312.

26. "In 1907, while I was writing a review of the consequences of special relativity, I realized that all natural phenomena could be discussed in terms of special relativity except for the law of gravitation. I felt a deep desire to understand the reason behind this." Ibid., 179.

27. Ibid., 178.

28. Ibid., 179.

29. Newton described gravity as a force, but physicists after Einstein describe it in terms of a gravitational field. Specifically, at each point in space they define the field as the force that would be experienced by a given mass, if it were positioned there.

30. Pais, *"Subtle Is the Lord—"*, 257.

31. North has observed that, in the fifty years ending in 1915, "at least eighty papers devoted to non-Euclidean statics, dynamics and kinematics" were published. *Measure of the Universe*, 73–74.

32. Specifically, the "parallel postulate" stated that if a straight line falling on two straight lines makes the facing interior angles less than two right angles, then those two straight lines, if extended indefinitely, meet on that side where the angles are less than two right angles. The postulate was reformulated several times; perhaps the clearest reformulation is "Playfair's axiom," after the English mathematician John Playfair (1748–1819). It says that, given a line and any point not on that line, there exists one and only one parallel to the given line passing through the given point.

33. The first was al-Gauhari, a contemporary of Muhammad ibn Musa al-Khwarizmi. The second was Abul Hasan Thabit ibn Qurrah, who, in the second half of the ninth century, wrote two accounts of the parallel postulate. In the early eleventh century a scholar named Abu Ali al-Hasan ibn al-Haytham addressed the problem in a work entitled *Commentary on the Premises to Euclid's the "Elements."* In the early twelfth century, Omar Khayyam, known in the West as a poet and the author of the *Rubáiyát*, devised what he believed to be proofs to the parallel postulate; and in the thirteenth century, the mathematician and philosopher Nasir Eddin al-Tusi did the same.

34. Meschkowski, *Evolution of Mathematical Thought*, 31.

35. Gray, *Ideas of Space*, 107.

36. Ibid., 86.

37. Ibid. The elder Bolyai had known Gauss at the University of Göttingen. He asked Gauss for an opinion of his son's work, and Gauss answered (rather uncharitably) that to praise János's work would be tantamount to praising himself, as he had held similar ideas for many years. Naturally, both Bolyais were disturbed by Gauss's response, János so much so that he published nothing more.

38. The work was translated into English by the mathematician William Clifford. In a paper read in 1870, he had imagined that "small portions of space are in fact of nature analogous to little hills on a surface which is on average flat; namely, that the ordinary laws of geometry are not valid in them." He suggested further that "this property of being curved or distorted is continually being passed on from one portion of space to another after the manner of a wave." Clifford had begun to discern the vague outline of something like general relativity. North, *Measure of the Universe*, 73.

39. For details of this interesting period in Einstein's work on general relativity, see Pais, *"Subtle Is the Lord—"*, 211–213.

40. The shift was 1.38 seconds of arc. Newton's laws explained 1.28 seconds of it but could not account for the remaining 0.10. This was an almost unimaginably small number (an angle about equal to that subtended by the width of a toothpick held at a distance of one kilometer), but through the late nineteenth and early twentieth centuries, astronomers had worked hard to explain it. Their hypotheses were diverse and demonstrated rich imaginations. They proposed a number of hypotheses: a planet inside Mercury's orbit, a swarm of asteroids, a Mercurian moon, and even interplanetary dust. All these ideas assumed that the cause of the perturbation was an undiscovered mass, and that Newton's law of gravitation was unimpeachable.

41. Whitehead, *Science and the Modern World*, 22.

42. Because light coming from the surface of the Sun has a longer wavelength than does light coming from the same material in Earth-based laboratories, natural "clocks" run more slowly when they are nearer gravitating masses. The demonstration that follows in the text is borrowed and adapted from Kip Thorne's rendering in Chapter 2 ("The Warping of Space and Time") of his *Black Holes and Time Warps: Einstein's Outrageous Legacy.*

43. In the 1950s, physicist Wolfgang Rindler began to call this surface a "horizon" because, like any horizon, it hides or eclipses a star. Of course a

black hole's horizon eclipses the star completely, along all possible sight lines. Rindler's coinage has since been expanded to "Schwarzschild horizon" or, more commonly "event horizon," with the word "event" used (as in Minkowski diagrams) to mean a point in spacetime. No signal from an event inside the black hole can ever reach an observer outside the event horizon, and no observer outside the event horizon can ever receive a signal from an event inside it. To forestall confusion, this history will use the term "event horizon" to describe the surface of a black hole, regardless of the period under discussion.

44. Misner, Thorne, and Wheeler, *Gravitation*.

45. Gold, "Rotating Neutron Stars," 731.

46. Gold, "Mother and Baby Paradox," 113.

47. In 1975, the same year that Gold published his thoughts on mass-shell time dilation, a group of Princeton physics graduate students noted that if a self-supporting shell of the strongest possible material were to avoid becoming a black hole, its radius would have to be 4 percent larger than the Schwarzschild radius, the measure smaller than which any star will collapse. Lightman et al., *Problem Book*.

Chapter Three: "Unphysical" Time Machines

1. Van Stockum was also a part-time inventor and encouraged his family toward playful experimentation of all sorts. Willem's elder sister, Hilda, gained fame as an author of children's books.

2. Van Stockum, "Soldier's Creed."

3. Van Stockum, "Gravitational Field," 135.

4. The first to note that Van Stockum's solution produced closed timelike curves may have been S. C. Maitra in "Stationary Dust-Filled Cosmological Solution with $\Lambda = 0$ and without Closed Timelike Lines," *Journal of Mathematical Physics* 7 (1966): 1025–1030.

5. Pais, *"Subtle Is the Lord—"*, 473.

6. Ibid., 462.

7. Einstein's assistant Ernst Gabor Strauss noted, "The one man who was, during the last years, certainly by far Einstein's best friend, and in some ways strangely resembled him most, was Kurt Gödel, the great logician." Wang, *Reflections*, 114.

8. Ibid., 31.

9. Ibid., 77.

10. Ibid., 115.

11. Ibid., 114–115.

12. Ibid., 32.

13. Ibid., 37.

14. Titled "A Remark about the Relationship between Relativity Theory and Idealistic Philosophy," Gödel's essay was published in 1949 in *Albert Einstein: Philosopher-Scientist*. In the same volume, Einstein responded, "Kurt Gödel's essay constitutes, in my opinion, an important contribution to the general theory of relativity, especially to the analysis of the concept of time. The problem here involved disturbed me already at the time of the building up of the general theory of relativity, without my having succeeded in clarifying it." 687. In the same year, Gödel published a more technical version of the position (Gödel, "Example") and made a presentation on the subject to an audience at Princeton (Gödel, "Lecture").

15. Gödel, *Collected Works*, 285.

16. Ibid.

17. Since about 1925, astronomers have known that the universe is expanding, and from any point in space it appears that galaxies are moving away from us. In fact, on average all the galaxies are moving away from each other, and there is no preferred point of view. The situation is best visualized with an oft-cited and homely comparison: The Milky Way Galaxy is like a raisin in a loaf of rising bread, with other galaxies represented by the other raisins, and space itself represented by the rising bread between the raisins. The other raisins seem to be moving away from us. But if we were on another raisin, we would also see all the other raisins moving away; the reason, of course, is that the space between the raisins is expanding.

18. Schilpp, *Albert Einstein*, 687–688.

19. Wang, 8–9.

20. Wheeler, *Geons*, 309–310. Kip Thorne's memory is that Wheeler had no particular reason for arranging the meeting, except that all were interested in the somewhat philosophical aspects of physics. (Author's interview with Thorne, August 18, 2005.)

21. Wheeler, 310.

22. The content and implications of the letters are discussed by Wang (212–218).

23. If Gödel considered the relation of the paradox to his groundbreaking theorem, or even considered the paradox in isolation, his papers make no mention of it.

24. Liversidge, "Frank Tipler."

25. Tipler, "Rotating Cylinders," 2205.

26. Tipler, "Causality Violation."

27. Gödel himself appreciated that his rotating universe might prove a physical impossibility for this reason alone. *Collected Works*, 285.

28. W. H. Williams, "The Einstein and the Eddington" (1924), excerpted in Chandrasekhar, *Eddington*, 41.

Chapter Four: Pastward Time Travel . . . Seriously

1. Einstein was at Caltech as a guest professor in 1930 and again from December 1931 to March 1932.

2. On the day that Thorne, Wheeler, and Misner visited Gödel, this was their work in progress.

3. Irion, "Kip Thorne," 1488.

4. Sagan, *Cosmic Connection*, 241.

5. Cooper, *Search for Life on Mars*, 70.

6. For a recent overview of the field, see Jeffrey Bennett, Seth Shostak, and Bruce Jakosky, *Life in the Universe* (San Francisco: Addison-Wesley, 2003).

7. Thorne's response to the challenge of Sagan's query is described in detail in Chapter 14 of Thorne's *Black Holes and Time Warps*, from which much of this chapter is derived.

8. Schwarzschild himself had not used the term, and over ensuing decades the phenomenon had been called by several names: "Schwarzschild singularity," "frozen star," and "collapsed star." Each had its moment, and its own advantages and disadvantages. The term "black hole" was coined by John Wheeler in 1967 and is now employed universally.

9. Landau, "Origin of Stellar Energy," 333.

10. Details of Landau's work are reported in Chapter 5 of Thorne's *Black Holes and Time Warps*. In addition to its other virtues, Thorne's book was the first to uncover the history of the work on black holes by Soviet physicists in the twentieth century, and to my knowledge it remains the only one to describe that history in detail.

11. Oppenheimer and Volkoff, "On Massive Neutron Cores," 540.

12. Oppenheimer and Snyder, "On Continued Gravitational Contraction."

13. What, then, of Einstein, whose theories of gravity warping spacetime started it all? He seems to have avoided the subject of black holes until 1939, when, reacting to mounting interest, he developed his own model. It was a spherical swarm of particles with different planes of revolution. As the sphere shrank, the particles would have to revolve faster and faster,

and near the horizon they would approach the speed of light. He concluded that further shrinkage would be impossible, but he seems not to have accounted for the loss of angular momentum through the ejection of mass. Mostly, though, Einstein by the late 1930s had begun his search for a unification of gravitation and electromagnetism. Black holes were nothing more than a possible consequence of general relativity, a theory that he regarded as provisional.

14. Hall, "Thorne's Time Machine," 71.

15. Boslough, *Masters of Time*, 216.

16. Current estimates of the threshold for gravitational collapse are nearer three solar masses. In fact, Wheeler was correct; the collapsing star did eject some mass.

17. Wheeler, *Geons*, 312.

18. The joke that a topologist is someone who does not know the difference between a coffee cup and a doughnut derives from the fact that one object can be transformed into the other without breaking or tearing its surface.

19. In 1969, British astrophysicist Donald Lynden-Bell would suggest that enormous black holes lived in the center of galaxies; and in 1971, Stephen Hawking would propose that small "primordial" black holes may have been created in the big bang, and some of them may have survived into the present.

20. Overbye, *Lonely Hearts*, 204.

21. Carter, "Complete Analytic Extension"; and Carter, "Global Structure."

22. Since about 2000, astronomers have discovered evidence that the visible universe is expanding, and that that expansion is accelerating far too fast for the mutual gravitational pull of its parts ever to slow and reverse it. The big crunch, it seems, has gone the way of the aether.

23. Abell, *Exploration of the Universe*, 594.

24. Wali, "Chandra," 271–272.

25. Thorne, *Black Holes and Time Warps*, 486.

26. This, too, was partly Wheeler's coinage: "There is a net flux of lines of force through what topologists would call a handle of the multiply-connected space and what physicists might perhaps be excused for more vividly terming a 'wormhole.'" Misner and Wheeler, "Classical Physics," 525.

27. By 1985 the depiction of learned men discussing and debating a scientific mystery was a formulaic but necessary staple of science fiction; all such scenes owed a literary debt to the first chapter of Wells's *The Time Machine*.

28. Morris and Thorne, "Wormholes in Spacetime," 406.

29. Planck length $= 1.6 \times 10^{-35}$ meters.

30. This "granular" concept of time had an interesting precedent. In the eighth century, al-Baqillani of Basra proposed an idea for time as utterly discrete moments. According to this idea, the world was created anew in each moment, and in each interval subsequent to that moment—an interval that passed so quickly as to be undetectable—the world was destroyed. Thus the world was continually being created, annihilated, and created again.

31. Morris and Thorne, 407.

32. Roman had been an undergraduate at the University of Maryland when he heard Frank Tipler speak about rotating cylinders. Tipler at the time was a graduate student at the same institution.

33. Author's interview with Roman, February 20, 2005.

34. Morris and Thorne, 407.

35. The observation is meant to demonstrate the speed at which technologies develop—that a fourteenth-century alchemist would regard a particle accelerator, existing a mere six hundred years later, as magic; and that a physicist from our time would regard a hypothetical technology six hundred years in our future as beyond his understanding. Of course, the observation is made from the point of view of an insufficiently advanced civilization. To this we might add that—again from our point of view—any sufficiently advanced civilization of extraterrestrial origin would be indistinguishable from any sufficiently advanced civilization from our own future.

36. Price and Thorne, "Membrane Paradigm," 72.

37. Thorne, *Black Holes and Time Warps*, 508.

38. Morris, Thorne, and Yurtsever, "Wormholes, Time Machines," 1446.

39. "Wormholes and Time Machines."

40. Hall, "Thorne's Time Machine," 70–71.

41. Irion, 1491.

42. Thorne, *Black Holes and Time Warps*, 468.

43. Thorne's own account of the foregoing episodes appears in Chapter 14 of *Black Holes and Time Warps*.

Chapter Five: Paradox

1. Quoted in Ferris, *Whole Shebang*, 336, n. 5.

2. Friedman, "Back to the Future," 305–306.

3. Thorne, *Black Holes and Time Warps*, 508.

4. Thorne, *Black Holes and Time Warps*, 278.

5. Novikov, *River of Time*, 106.

6. Thorne had known that he wanted to study physics, and he had an intuition (which would prove correct) that many interesting discoveries would be made by Soviet physicists. (Author's interview with Thorne, August 18, 2005.)

7. Novikov, *River of Time*, 108.

8. Morris, Thorne, and Yurtsever, "Wormholes, Time Machines" 1448.

9. Browne, "3 Scientists Say."

10. Novikov, *Evolution of the Universe*, 169. Thorne does, however, recall discussing the matter for the first time with Novikov in Moscow sometime in the late 1980s, and Novikov's reminding him of the suggestion in his book. (Author's interview with Thorne, August 18, 2005.)

11. Escher, *Escher on Escher*, 78.

12. L. S. Penrose and Penrose, "Impossible Objects," 31.

13. This particular paradox is alluded to so frequently that it is somewhat startling to consider that there actually *was* a story that first presented it in print: according to John L. Flynn (in "Time Travel Literature"), that work was Nathan Schachner's 1933 "Ancestral Voices."

14. Mainstream cinema of the last twenty years or so has explored the dramatic possibilities offered by temporal paradoxes. Well-known examples are *Back to the Future* (1985) and its sequels, *Bill & Ted's Excellent Adventure* (1989), and *The Terminator* (1984) and its sequels.

15. Lewis, "Paradoxes," 150.

16. From Arthur C. Clarke's 1950 story "Time's Arrow," in *Collected Stories*.

17. In order of mention: H. L. Gold's "The Biography Project" (1955), D. Franson's "On Time in Alexandria" (1980), Michael Moorcock's *Behold the Man* (1969), and Harry Harrison's *The Technicolor Time Machine* (1967).

18. This is, of course, *Peabody's Improbable History*. The dog was "Mr. Peabody," the boy was "Sherman," and the time travel was accomplished through the "Wayback Machine." In various formats the series ran for ninety-one episodes from 1959 to 1964.

19. Nahin, *Time Machines*, 321.

20. Heinlein, *Fantasies*, 348.

21. In fact, Heinlein delineates his narrator's involuted worldline so well that physicist Michio Kaku was able to illustrate it in his 1994 book *Hyperspace* (p. 241).

22. Eddington, *Nature of the Physical World*, 293.

23. Einstein, "Religion and Science."

24. Hawking and Ellis, *Large Scale Structure*, 189.

25. For more on this, see Novikov's own exposition of his idea in Hawking et al., *Future of Spacetime*.

26. Echeverria, Klinkhammer, and Thorne, "Billiard Balls," 1078.

27. Author's interview with Thorne, August 18, 2005.

28. The work was Robert Forward's *Timemaster* (New York: TOR Books, 1992), which is in turn cited in Echeverria, Klinkhammer, and Thorne, "Billiard Balls."

29. In the usual manner of things, authors of science fiction stories develop plots from established facts or speculation based upon such facts. The science *precedes* the fiction, and the science fiction follows the science. The rich corpus of science fiction that involves time travel is perhaps among the reasons that the subject has about it an air of being unsuited for serious study. So there was no small irony in the fact that the most sustained and intense research in the field was generated by the innocent inquiry of a first-time author of science fiction, and that the further work described here was provoked by an idea from notes for a science fiction novel. It seems fitting that the cause and effect we expect are reversed or undone by time travel fiction.

30. Feynman, Leighton, and Sands, *Feynman Lectures*, Vol. 1, ch. 37, p. 1.

31. Einstein was perhaps the most articulate critic of quantum mechanics, and much of his distaste sprang from interpretations that suggested "nonlocality"—the interaction between phenomena widely separated in space. He called it, rather dismissively, *spukhafte Fernwirkungen*—"spooky action at a distance." In 1935, the same year that he and Nathan Rosen had discovered the two possible Schwarzschild solutions to the black hole, they published a paper with colleague Boris Podolsky (Einstein, Podolsky, and Rosen, "Can Quantum-Mechanical Description"). It enlarged a system much like the double-slit apparatus to enormous scale, intending to expose the peculiarities of the Copenhagen interpretation (which will be discussed shortly in the text).

32. Since the 1920s, physicists have had reason to believe that electrons sent through the apparatus one at a time would also produce the pattern, and in the 1980s and 1990s, two research groups confirmed these suspicions.

33. Gribbin, *Schrödinger's Kittens*, 8.

34. Einstein seems to have used the phrase for the first time in a letter to Max Born dated December 4, 1926. He wrote, "Quantum mechanics is very impressive. But an inner voice tells me that it is not the real thing. The the-

ory produces a good deal but hardly brings us closer to the secret of the Old One. I am at all events convinced that *He* does not play dice." (Born, *Born-Einstein Letters*, 91.) Einstein's mentions of a deity, it should be noted, refer not to a personal God concerned with the fate of human beings, but to what he called "Spinoza's God," revealed in the harmony of all that exists. In his autobiography, Einstein writes that his more traditional religious inclinations were short-lived, ending at age twelve. For a thorough discussion of this subject, see Jemmin, *Einstein and Religion*.

35. Wheeler, *Geons*, 168.

36. "The sum over histories meshes especially nicely with the principle of self-consistency. To impose that principle one need only ensure that the sum includes every self-consistent history, and only self-consistent ones. . . . [We] have found that [this approach] gives a unique, self-consistent set of probabilities for the outcomes of all sets of measurements one might imagine making; i.e., it removes the classical theory's multiplicity of solutions." Friedman et al., "Cauchy Problem," 1926.

37. Novikov, *River of Time*, 109.

38. In December 1991 it was given back its older name: the Russian Academy of Sciences.

39. Frolov and Novikov, "Physical Effects," 1062.

40. Ibid., 1058.

Chapter Six: Cosmic Strings and Chronology Protection

1. Chandrasekhar was there as well, and he had been since 1937. By 1988 he was in his late seventies and still producing important work. Only five years earlier, the same year that he received the Nobel Prize, he had overseen the publication of his work *The Mathematical Theory of Black Holes*.

2. Geroch, "Topology."

3. Von Baeyer, *Emergence*, 116.

4. Electromagnetic vacuum fluctuations may be thought of as electromagnetic waves or as photons, and gravitational fluctuations may be thought of as gravitational waves or gravitons. There is one way, though, in which the phenomena are very different. Electromagnetic vacuum fluctuations are an aspect of everyday life: turning on a fluorescent light initiates a process involving random electromagnetic fluctuations. But gravitational waves and gravitons are yet to be observed.

5. Hiscock and Konkowski, "Quantum Vacuum Energy."

6. Recall Gott's hypothetical mass shell from Chapter 2.

7. Tryon, "Is the Universe," 396.

8. In 2005, a U.S.-Ukrainian team of astronomers based at the Harvard-Smithsonian suggested that a double quasar near the Big Dipper is actually two images of a single quasar split by a cosmic string. At about the same time a Russian-Italian team at Federico II University in Naples put forth a still better candidate: two mirror images of a galaxy. Neither is definitive, and Gott believes that each phenomenon may be explained by more familiar causes. (Author's interview with Gott, November 11, 2005.)

9. Cutler, "Global Structure."

10. Lemonick, "How to Go Back," 74.

11. Hawking, "Chronology Protection Conjecture," in *Sixth Marcel Grossman Meeting.*

12. Hawking, "Chronology Protection Conjecture," *Physical Review D,* 610. Hawking's comment has been much quoted. Science fiction authors since the 1940s had been suggesting that time travel offered possibilities to the tourism industry. Robert Silverberg's 1969 novel *Up the Line* in particular raised the specter of tourists mobbing famous moments in history.

13. Hawking writes, "It is worth remarking that, even if we could distort the light cones in the manner of this example, it would not enable us to travel back in time to before the initial surface S. That part of the history of the universe is already fixed. Any time travel would have to be confined to the future of S." "Chronology Protection Conjecture," *Physical Review D,* 605.

14. Recall Wells's "Very Young Man," who realizes that time travel might offer opportunities for financial gain. Just two years before Thorne and Morris published their seminal paper, M. R. Reiganum observed that a time traveler from our future would know which financial markets would increase in value, and so which were worthy of investment. If enough time travelers were making such investments, the long-term consequence would be to drive interest rates to zero. His conclusion, offered with tongue at least partly in cheek, was this: because interest rates are not zero, time travel is demonstrably impossible. Reiganum, "Is Time Travel Possible?"

Chapter Seven: Novikov's Wild Ideas

1. Novikov, "Time Machine."

2. The paper under discussion uses the terms "jinn" and "jinni" interchangeably. We adopt the shorter version in the interests of economy.

3. It was indeed an interesting worldline. Gamow had been a classmate of Lev Landau, and he was profoundly disturbed by the Stalinist regime of the Soviet Union. When in 1933 he was allowed to attend a professional conference in Brussels, he did not return. A year later he accepted a posi-

tion at the George Washington University in Washington, DC. Gamow went on to make significant contributions to quantum physics, to argue persuasively in favor of the big bang model of the origin of the universe, and to propose correctly that the DNA molecule could carry a coded message. He was also a talented and accomplished writer of science for lay audiences. His best-known work of this type is his 1940 *Mr. Tompkins in Wonderland*, which inspired a young Roger Penrose to a embark upon a career in physics. As a child, Thorne had read Gamow's *One, Two, Three . . . Infinity* three times; when he mentioned that fact to Gamow in a letter, the elder physicist sent him a Turkish edition, inscribed "To my dear colleague Kip Thorne, so that he will not be able to reread it a fourth time!" Hall, "Thorne's Time Machine," 71.

4. Many science fiction stories concern artifacts arrived from the future, always with interesting consequences. Notable examples are "The Twonky" (1942) and "Mimsy Were the Borogoves" (1943)—both by Henry Kuttner and C. L. Moore, writing under the pen name Lewis Padgett—and "The Little Black Bag" by C. M. Kornbluth. Jeannot Szwarc's 1980 film *Somewhere in Time*, in a nice bit of metonymy, features a jinn watch.

5. Eddington, *Nature of the Physical World*, 74.

6. Escher, *Escher on Escher*, 79.

7. As a civilization, we have practice in devising messages that endure for long periods, and messages that are directed across many generations. Time capsules are a means for a society from one time to explain itself to a society from a later period, and to do so in its own "voice," without intermediaries or interpreters like historians and journalists. The mission of a time capsule may be judged successful when it has transmitted its contents to a later generation. Transmitting the *meaning* of those contents would be somewhat more challenging because it would require a message that transcended language and culture. But even the transmission of material with no accompanying interpretation can work only if members of a community or institution remember that the capsule exists, and leave it undisturbed until the date specified on the capsule and or in a marker set near it. It is interesting to note that Einstein was asked to compose a message to be placed in a time capsule that would be sealed at the 1939 World's Fair and opened in the year 6369. (The text of the letter is in Albert Einstein, *Ideas and Opinions*. New translations and revisions by Sonja Bargmann. [New York: Dell Publishing, 1973]. Other details are recounted in the *New York Times*, September 16, 1938.)

8. Similar schemes have been acted upon more recently. The city of Perth, in Western Australia, declared March 31, 2005, "Destination Day" for time travelers; and on May 7, 2005, students at MIT convened a "Time Trav-

eler's Convention." To the best of my knowledge no guests from the future were seen at either event. I can think of no reason not to make another offer here. So, IF TIME TRAVEL IS POSSIBLE, PLEASE CONTACT THE AUTHOR UPON THE PUBLICATION DATE OF THIS BOOK. (The reader is promised a full account of any and all responses in any subsequent edition of this work.)

9. Thorne, *Black Holes and Time Warps*, 486.

10. In 1994, two years after the Aspen workshop that is described in Chapter 8, NASA sponsored a workshop designed to provide a forum for ideas of faster-than-light travel and communication. One outcome was a paper by John Cramer and several other physicists (among them Mike Morris and Matt Visser) that suggested a means by which wormholes might be detected. (Cramer et al., "Natural Wormholes.") It rested upon the supposition that over time the "exit" mouth of a wormhole would acquire negative mass, and that such mass would generate a repulsive gravitational field that would in turn produce a distinctive "gravitational lensing" effect—a ring of light. Astronomers routinely monitor broad swaths of sky for dark-matter objects—that is, objects that are invisible to our telescopes yet generate powerful gravitational fields—and they have found several. None of their signatures match what Cramer and his colleagues expect would be produced by a wormhole.

11. As of this writing, the spacecraft that has traveled farthest from Earth is *Voyager 1*. Launched in September 1977, *Voyager 1* (as of the time of the book's publication) is about 12 light-hours (or 7.8 billion miles) from Earth, well beyond the orbit of Pluto. On interstellar scales, 12 light-hours is a rather modest distance. The nearest star is 4.3 light-*years* away, and *Voyager 1* (traveling at a rate of 39,000 miles an hour, or about one ten-thousandth the speed of light) will take about 40,000 years to traverse the equivalent distance. The spacecraft has already outlived its design life span by several years, and its radioisotope electrical generators should continue to supply it with power until at least the year 2020. Thus we may predict with some certainty that the total active life span of humanity's first interstellar spacecraft will be roughly forty-three years.

12. Many wormholes in television and film are inaccurately depicted as a tornado-like funnel; this error may stem from the mistaken reading of an embedding diagram as a naturalistic rendering of three-dimensional space.

13. "The Time Lords," part of the BBC's *Horizon* series, aired December 2, 1996. It subsequently appeared in the United States as a *NOVA* special titled "Time Travel" and aired by PBS on October 12, 1999.

14. Suppose the spacecraft left Earth in the year 2100. The voyage from Earth to the present mouth of the wormhole took 500,000 years. In other

words, the spacecraft arrived in the year 502100. This means that, if our spacecraft is to keep its appointment back in 2100 with the advanced computer and the automated plant, it will first need to travel back in time 1,000,000 years—at least 500,000 years to return to the time it left Earth, and another 500,000 years to give itself the additional time it will need for the return journey to Earth. If the difference between the mouths were exactly 1,000,000 years, and the past and present mouths were the same distance from Earth, the course of action would be fairly straightforward. The spacecraft would begin its return to Earth immediately and expect to arrive in 2100. But in all likelihood the wormhole was not designed with our spacecraft in mind. Suppose that the temporal difference between the mouths is not 1,000,000 years, but only 10,000 years, meaning that our spacecraft finds itself at a moment 990,000 years later than its target date. It calculates that traveling back in time the remaining 990,000 years will require 999 trips through the wormhole. In normal space the present mouth is relatively nearby. Still, it requires 500 years of travel through normal space. The spacecraft will need to repeat this journey through normal space 999 times as well. In fact, total travel time in real space from past mouth to present will require 499,500 years. But it has, as it were, all the time in the world.

15. Novikov's 1998 book *The River of Time* elaborated on the problems of causal loops.

Chapter Eight: A Time Travel Summit

1. Penrose's expression acknowledges and plays off the mildly ribald connotations of Wheeler's original phrase "black hole" and "hair," the secondary meanings of which, Thorne says, Wheeler was surely aware. (Thorne, *Black Holes and Time Warps*, 257.) Wheeler himself suggests otherwise. (*Geons*, 296–297.) In 1992, Lawrence Ford and Thomas Roman performed an amusing riff on the phrase, naming a process by which the production of a singularity would render it momentarily naked as "cosmic flashing."

2. See "Acknowledgments" in Thorne, "Closed Timelike Curves," 314.

3. Morris and Thorne, "Wormholes in Spacetime," 407.

4. Visser, "Traversable Wormholes: Some Simple Examples," 3183.

5. Readers who find this description familiar will recall that one of the tablet-shaped "monoliths" from Arthur C. Clarke's 1968 work *2001: A Space Odyssey* operated as a wormhole mouth, and its shape was like those described in Visser's paper. Its edges were the exact ratios of one to four to nine—the squares of the first three integers—engineered to a precision far beyond human capabilities. Whether such perfection is merely aes-

thetic, a passive display of power, or something else, Clarke's human characters can only guess.

6. In the early 1980s, five years before Morris and Thorne's teaching paper, several physicists had made detailed predictions of what might happen to wormholes subjected to inflation: Katsuhiko Sato, Misao Sasaki, Hideo Kodama, and Kei-ichi Maeda, "Creation of Wormholes by First Order Phase Transition of a Vacuum in the Early Universe," *Progress of Theoretical Physics* 65 (1981): 1443–1446. Roman's calculation, not drawn from their work, was for a wormhole in "de Sitter spacetime"—that is, a three-dimensional universe beginning in the infinite past with infinite size, then undergoing contraction followed by expansion.

7. Roman, "Inflating Lorentzian Wormholes."

8. In 1971, Zel'dovich proposed that the swirl of space outside the event horizon caused rotating black holes to radiate by a process best understood not through general relativity but through the (poorly understood) laws of quantum gravity.

9. Guth and Farhi had imagined a similar result for their "universe-in-a-lab." To those viewing the experiment from the parent universe, it would be as though they had created a miniature, and very short-lived, black hole. Because the demise of a miniature black hole would be rather spectacular, so would the birth of a universe. At the moment of disconnect, the parent universe would experience the energy equivalent of a five-hundred-kiloton nuclear explosion.

10. Visser, *Lorentzian Wormholes*, 174.

11. Ford and Roman, "Quantum Interest Conjecture."

Chapter Nine: A Parade of Time Machines

1. Allen Everett would later term it the "banana peel" mechanism.

2. Visser, "From Wormhole to Time Machine," 557.

3. A "Hausdorff space" is a topological space in which points can be separated by regions called "neighborhoods." A "differentiable" surface is a surface on which one can perform calculus. A "manifold" is an abstract space in which every point has a neighborhood that resembles simple Euclidean space but may have a more complicated global structure. Manifolds may exist in any dimension. In a one-dimensional manifold every point has a neighborhood that looks like a segment of a line. Lines and circles, for instance, are one-dimensional manifolds. In a two-dimensional manifold, every point has a neighborhood that looks like a disk. Planes, the surfaces of spheres, and the surfaces of tori are all two-dimensional manifolds. So a four-dimensional "Hausdorff differentiable manifold" is a

four-dimensional space in which points can be separated by neighborhoods and on which one can perform calculus.

4. Visser, "From Wormhole to Time Machine," 558.

5. Peter Damian, *Letters 91–120*, 356.

6. The idea had already enjoyed a long history, rooted among other places in a verse from Psalms: "A thousand years in your sight are as yesterday when it is past" (Ps 90:4).

7. Peter Damian, 358.

8. Visser, "Wormholes, Warpdrives."

9. In his best-selling 1988 work *A Brief History of Time*, Hawking counted three arrows: the thermodynamic arrow, the direction in which disorder or entropy increases; the psychological arrow, the direction in which we sense that time passes; and the cosmological arrow, the direction in which the universe is expanding. Penrose's earlier list included four more: quantum mechanical observations, the retardation of radiation, black holes versus white holes, and the decay of the K^0 minus meson discussed in this chapter.

10. "The conventional wisdom has it that, despite an initial appearance to the contrary, the framework of quantum mechanics contains no arrow." Roger Penrose, "Singularities," 582.

11. Ibid., 583.

12. Visser, "Wormholes, Baby Universes," 1123.

13. Visser, "From Wormhole to Time Machine."

14. The overview that follows in the text is an attempt to represent the range of approaches to the question that physicists undertook in those years and is by no means exhaustive. In the years between 1990 and 1999, *Physical Review D* alone published fifty articles whose subject directly involved closed timelike curves.

15. For a vivid description, see Gott, *Time Travel*, 137–138.

16. Lyutikov, "Vacuum Polarization"; and Visser, "Van Vleck Determinants."

17. Alcubierre, "Warp Drive."

18. Allen E. Everett, "Warp Drive" (submitted for review in 1995).

19. Pfenning and Ford, "Unphysical Nature."

20. The work described in the remainder of this paragraph appeared in a preprint in 1995 and was published in *Physical Review D* in 1998.

21. Allen E. Everett and Roman, "Superluminal Subway."

22. Li, Xu, and Liu, "Complex Geometry."

23. Li, "New Light."

24. Li, "Must Time Machines Be Unstable?"

25. Feynman, Leighton, and Sands, *Feynman Lectures*, Vol. 1, ch. 19, p. 1.

26. Carlini et al., "Time Machines."

27. Li-Xin Li, for instance, begins a paper, "Among the difficulties in building time machines . . ." (Li, "New Light," R6037); and Cassidy and Hawking admit, "It was hoped that the back reaction would be sufficiently strong enough to prevent the formation of CTCs" (Cassidy and Hawking, "Models," 2372).

28. Visser, "Traversable Wormholes: The Roman Ring."

29. Cassidy and Hawking, "Models."

30. Li, "Time Machines."

31. Kay, Radzikowski, and Wald, "Quantum Field Theory."

32. The situation was roughly similar to one in astronomy: it had long been a standing tenet that the horizon of a black hole, if it could somehow be examined closely, might be a window to physics that would involve Planck-scale fluctuations in the geometry of space, but there was little expectation that anyone could get close enough to conduct such an examination.

Chapter Ten: Time Travel in Another Kind of Spacetime

1. Borges, *Ficciones*, 100.

2. Wheeler, *Geons*, 291.

3. Shikhovtsev and Ford, "Biographical Sketch."

4. Paul Davies, *Other Worlds*, 137.

5. Jammer, *Philosophy of Quantum Mechanics*, 278. Pearle was alluding to the principle of theoretical parsimony put forth by William of Ockham in the fourteenth century. It may be a useful logical tool, but it may at the same time be a misleading predictor of nature, which—as poppy seeds and stars show us—can be quite profligate.

6. Folger, 7.

7. P. C. W. Davies and Brown, *Ghost in the Atom*, 92–93.

8. This idea, too, was anticipated by science fiction. P. Schuyler Miller's 1944 "As Never Was" (published in January of that year in the magazine *Astounding Science-Fiction*) includes the following passage: "As any schoolchild learns, the time shuttler who goes into the past introduces an alien variable into the spacio-temporal matrix at the instant when he emerges.

The time stream forks, an alternative universe is born in which his visit is given its proper place, and when he returns it will be to a future level in the new world which he has created. His own universe is forever barred from him." Science fiction stories set in a world whose history has diverged from our own are termed "alternate histories" and represent a large subgenre of science fiction. Jorge Luis Borges's 1941 short story "The Garden of Forking Paths" includes haunting meditations on alternate existences. The narrator is told, "Differing from Newton and Schopenhauer, your ancestor did not think of time as absolute and uniform. He believed in an infinite series of times, in a dizzily growing, ever spreading network of diverging, converging and parallel times. This web of time—the strands of which approach one another, bifurcate, intersect or ignore each other through the centuries—embraces *every* possibility." (*Ficciones*, 100.)

9. Deutsch describes this idea in an interview recorded in P. C. W. Davies and Brown, *Ghost in the Atom*.

10. Deutsch, *Fabric of Reality*, 310–311.

11. Pagels, *Cosmic Code*, 112.

Chapter Eleven: Explaining the Apparent Absence of Time Travelers

1. Visser, *Lorentzian Wormholes*, 261.

2. Hugh Everett III died of a heart attack in 1982.

3. Allen E. Everett, "Time Travel Paradoxes." In the same paper, Everett offered a neat counterexample to Novikov's self-consistency principle. "Suppose we place at the early time mouth of the wormhole a device to detect [a] BB if it emerges. (One might, e.g., have a spherical grid of current carrying wires enclosing the wormhole mouth . . . thin enough to be broken by the BB with its given speed and spaced closely enough that a BB cannot emerge from the wormhole without breaking at least one of the wires.) Suppose further that we connect the detector in such a way that, if a BB is detected, a signal is sent at light speed activating a mechanically operated shutter, which deflects the incident ball at some later point in its path so that it does not enter the late-time wormhole mouth. One can include a requirement that the signal is sent only if a BB emerges from the wormhole before the incident ball reaches the shutter . . . [I]t is difficult to see how there can be any self-consistent, and hence physically acceptable, solution. Thus we seem to be back to the grandfather paradox in the form of a BB which enters the wormhole if and only if it does not enter the wormhole." (Ibid., 2.)

4. Author's interview with Allen Everett, April 20, 2006.

5. "The Time Lords," BBC *Horizon* series, aired December 2, 1996; on October 12, 1999, PBS aired it as a *NOVA* program called "Time Travel."

6. "Sagan on Time Travel" in "Time Travel," *NOVA Online*, November 2000, www.pbs.org/wgbh/nova/time/sagan.html.

7. The conference was called "Where Are They? A Symposium on the Implications of Our Failure to Observe Extraterrestrials." According to the most reliable accounts, in 1943 during a social gathering at Los Alamos, the physicist Enrico Fermi, apropos of nothing, asked, "Where are they?" When the others responded, "Where are who?" Fermi replied "Why, the extraterrestrials."

8. In 1980, Frank Tipler both took up this argument and furthered it. He suggested that, even if extraterrestrials were not here themselves, surely they would have by now populated the galaxy with their representatives— intelligent, self-replicating machines. Even given the most conservative estimates of the number of civilizations, one might expect at least a few to have traveled those distances and left evidence—perhaps detectable radiation from a drive, or an artifact on Earth or the Moon. To a person, researchers agree that there is no such evidence. (SETI researchers discount the claims of authors like Whitley Strieber and Erich von Däniken as unable to withstand rigorous proofs.)

9. Ball, "Extraterrestrial Intelligence."

10. Deutsch, "Quantum Mechanics," 3211.

11. In Gödel's rotating universe, of course, one might travel pastward as far as one liked.

12. Michael, "Proposed Studies," 215.

13. To this, Arthur C. Clarke, on the record as doubting the possibility of time travel, has a specific rebuttal. He argues that however well intentioned and advanced a given society might be, it will inevitably harbor outlaws.

14. John D. Barrow, *Constants of Nature*, 170.

Chapter Twelve: Time Machines at the Ends of Time

1. Stapledon, *Last and First Men*, 181.

2. In what may be the first description of a self-consistent temporal loop, Stapledon's narrator explains, "It is true that past events are what they are, irrevocably; but in certain cases some feature of a past event may depend on an event in the far future. The past event would never have been as it actually was . . . if there had not been going to be a certain future event." (Ibid., 238.)

3. Russell, *Why I Am Not a Christian*, 106–107.

4. Freeman J. Dyson, "Time without End," 449.

5. Dyson *could* imagine complications. All organisms, even very cold ones, would overheat if they did not generate waste heat away from themselves. The problem is that the efficiency of any such radiator drops off much faster than the metabolic rate. Dyson conjectured that, for this reason, overheating would become a real danger, but he further conjectured that it might be avoided if the organisms lowered their average metabolic rate, and that they might do so by spending a large part of their lives in hibernation. Ibid., 453–457.

6. According to Tipler, the colonization of space that would make that heaven possible would be carried out not with manned spacecraft, an endeavor that would be enormously expensive, but with self-replicating spacecraft called Von Neumann probes, after a hypothetical concept based on the work of mathematician and physicist John von Neumann. Tipler's versions would be small and lightweight (perhaps weighing no more than a hundred grams), constructed with nanotechnology, and propelled by an antimatter drive. They would carry in their memory DNA sequences of humans and other terrestrial life. The first such probe, traveling at only one-tenth the speed of light, would be sent to a nearby star system. Once there, it would use the DNA sequences to create living cells that would develop into humans, our indirect descendants. Then, using locally available materials, it would make other probes. These would depart for other systems, each at one-tenth the speed of light, and the seeding process would continue. Some of our descendants would develop civilizations on planets; others would prefer to build habitats in space. By this means the volume of colonized space would increase in diameter by ten light-years in every sixty years. Because our galaxy is roughly 100,000 light years in diameter, it would be colonized completely in a mere 600,000 years. By the same process, in some 10^{19} years, the volume of colonized space would encompass the entire universe. This was a particularly timely moment because it was, by Tipler's calculations, about the time that the universe's expansion would slow, cease altogether, and reverse itself.

   The colonizers could not hope to reverse the collapse, but because its representatives would be nearly everywhere, they could exert some control over it. Tipler imagined that, with sets of strategically placed explosions, the suitably advanced civilization might speed parts of the collapse, thereby creating temperature differentials in adjacent areas, and great reservoirs of potential energy. Tipler believed that a civilization that survived so long would regard the conditions under which its ancestors lived as intolerable; it would also be highly principled. Consequently, what it would want to do with this energy, and what it *would* do, would be to make replicas of everyone who ever lived, along with their memories. To protect these beings from disorientation and/or shock, it would create for them an environment like the one they had known, but far more pleasant.

Here is where eschatology meets astrophysics, with rather spectacular results. Tipler called this reconstruction a "resurrection" and described the environment in which it would occur as "an abode which is in all essentials the Judeo-Christian Heaven." One of those essentials, of course, is immortality, and within the long but finite interval of the universe's collapse, Tipler's civilization manages even this. During the collapse enough energy would be available for an infinite amount of what a computer scientist would call information processing, and what the rest of us would call thinking or consciousness. A number—say the number 2—may be parsed into a finite series: $1 + 1$. It may also be parsed into an infinite series: $1 + \frac{1}{2} + \frac{1}{4} \ldots$ Likewise, this "subjective immortality" would be accomplished by an infinite parsing within a finite interval. So although the collapse would have a finite duration and in the big crunch an unequivocal and irreversible end, those who were reconstructed within the collapse would enjoy life everlasting, or the subjective experience of it.

7. Johnson, "Physics of Immortality."

8. Thorne and Tipler, "Cosmological Dialogue," 67.

9. It is interesting that the civilizations of both Dyson and Tipler had achieved a sort of immortality by tampering with the subjective experience of passing time—Dyson stretching it, Tipler squeezing it.

10. Tipler, *Physics of Immortality*, 1.

11. In his 1980 science fiction novel *Dragon's Egg*, Robert Forward imagined organisms living on the surface of a neutron star. The rate at which such organisms lived and developed would, by our standards, be greatly accelerated. They would be dense and microscopic, and their biochemistry would derive its energy not from electromagnetic forces but from the strong nuclear force. Forward described them in greater detail in a 1987 paper called "When You Live upon a Star."

12. Adams and Laughlin suggest that, because the era of black holes is extensive, lasting until the universe is 10^{100} years old, life near an event horizon would have a long time to take hold, and a very long time to evolve into complex forms. However, the sparse availability of energy would limit the complexity of any organism, and they doubt that intelligence could evolve there. (*Five Ages of the Universe.*)

13. William Shakespeare, *Henry V*, act 4, scene 1 (chorus).

14. Rees, *Our Cosmic Habitat*, 119.

15. Many, of course, have opted for the miracle—better known as argument by design. It has a long history. Early in his career, Kant wondered why space should have three dimensions—and not, for instance, four. His

answer was that, in four-dimensional space, the force of gravity would vary not as the inverse square of distance but as the inverse *cube* of distance, and planets would either fall into the Sun or spin out of their orbits. Theologian William Paley found evidence of intelligent design in much of the natural world. In the 1852 *Bridgewater Treatises*, William Prout observed that the fact that, unlike most liquids, water becomes less dense as it freezes and floats means that Earth's oceans in all seasons remain mostly liquid, and thus life is possible. Kidd et al., *Bridgewater Treatises*.

16. Hoyle, "Universe," 12.

17. Carter, "Large Number Coincidences," 291. Carter earlier made known some of the ideas in this essay in an unpublished 1968 Cambridge University preprint and at the Clifford Memorial Meeting in Princeton in 1970.

18. Leslie, *Universes*, 13.

19. Smolin, *Three Roads*, 202.

20. Harrison, "Natural Selection."

21. Nonetheless, the idea does not please those who would wish nature to be economical. If one civilization in every galaxy creates a universe, then there are billions of universes, all with a hundred billion galaxies. If one civilization in each of those . . .

     One is reminded of an essay by Loren Eiseley in his 1971 work *The Night Country*. Eiseley carries seeds in his pocket as he walks, and he imagines that to them he is nothing more than a means of conveyance and of reproduction: "'I have carried such seeds up the sheer walls of mesas and I have never had illusions that I was any different to them from a grizzly's back or a puma's paw" (p. 69). Likewise perhaps, the purpose of the suitably advanced civilization is to carry the seed of new universes.

22. Einstein, "Physics and Reality," 350. (This is often misquoted as "What is incomprehensible about the universe is that it is so comprehensible.")

23. John D. Barrow argues that, if a universe begins with its constants set far from the values that allow complexity (and life and us) to develop, those universes and their offspring can never produce the conscious beings needed to fine-tune the constants. (*Constants of Nature*, 286.)

24. Hawking, *Brief History of Time*, 141.

25. Gott and Li describe their challenges to Hawking and Hartle's reasoning in Gott and Li, "Can the Universe"; and Gott, *Time Travel*, 180–186.

26. Gott, 189.

27. Gott and Li, 36.

28. Again, a provocative speculation from a physicist has an antecedent in science fiction. In Ben Bova's 1984 novel *Orion*, humanity uses time travel to create itself.

Epilogue

1. Visser, Kar, and Dadhich, "Traversable Wormholes."

2. Roman, "Some Thoughts."

3. Ori, "Class of Time-Machine Solutions."

4. Hawking and Thorne have made a number of wagers concerning discoveries in physics. The index in Thorne's work *Black Holes and Time Warps* includes the entry "bets by Thorne," and five subentries beneath, one on black holes. The fifth entry reads, "no bet on time machines."

5. Lisa Dyson, "Chronology Protection."

6. Semeniuk, "No Going Back," 3.

7. Friedman et al., "Cauchy Problem."

GLOSSARY

absolute horizon See *event horizon*.

aether The hypothetical medium that was thought by nineteenth-century naturalists to pervade space.

Alcubierre warp drive A bubble of spacetime that is moved at supraluminary speeds.

anthropic principle An approach to cosmology that constrains fundamental natural constants and other circumstances by demonstrating that, were they otherwise, the universe would not support life and we would not be here to observe it.

arrow of time The direction of the "flow" of time that distinguishes past from future; more technically, apparent past–future asymmetry.

astrophysicist A scientist who applies physics to cosmic objects.

astrophysics The branch of physics involving cosmic objects and the natural laws that govern them.

averaged weak energy condition The "rule" that no observer measures negative energy when that energy is time-averaged over a worldline.

black hole According to classical physics, a region of space in which gravity is so strong that nothing, not even light, may escape. In 1974, Stephen Hawking somewhat complicated this definition by using quantum theory to show that small black holes (much smaller than those formed by the collapse of stars) emit radiation. A Schwarzschild black hole does not rotate and has no charge. A Reissner-Nordström black hole does not rotate but is electrically charged. A Kerr black hole, thought to be the most realistic, rotates and is expected to be electrically neutral.

block universe A picture of the universe, produced by the tenseless theory of time, in which past, present, and future are imagined to be held within it, as though embedded in a single, four-dimensional block.

bootstrap paradoxes Any of a variety of conditions that, by way of future-ward and pastward time travel, in effect become their own cause and may produce a jinn. One of two large classes of time travel paradoxes. Compare *grandfather paradoxes*.

Cauchy horizon The outer edge of a region of spacetime influenced by a given event. See also *chronology horizon*.

causality The doctrine that every new state must have resulted from an earlier state.

causal loop See *closed timelike curve*.

child universe A universe that might be produced artificially as a consequence of the creation of a false vacuum, and that might retain the physical constants of its "parent" universe.

chronology horizon A type of Cauchy horizon. Specifically, the boundary between a region of space that contains closed timelike curves and a region of space that has no closed timelike curves. Or the boundary separating a region in which time travel is possible from a region in which it is not possible.

chronology protection conjecture The speculation, first put forth by Stephen Hawking in 1991, that the laws of physics do not allow the creation of time machines.

classical physics The laws of physics that govern macroscopic objects; Newtonian physics and relativity; non-quantum mechanical physics. Compare *quantum mechanics*.

closed null curve A path in spacetime along which a light ray returns to its starting point in both space and time. Compare *closed timelike curve*.

closed timelike curve A path in spacetime along which a material particle returns to its starting point in both space and time. Less commonly, *closed timelike loop* or *causal loop*. Compare *closed null curve*.

closed universe A universe in which gravity is strong enough eventually to reverse its expansion. Compare *open universe*.

conjecture A judgment or educated guess made with incomplete or inconclusive information.

Copenhagen interpretation The interpretation of quantum mechanics associated with Niels Bohr and widely accepted as the conventional formulation. It asserts that a quantum system has no definite state until it is measured.

cosmic censorship conjecture The speculation, posed by Roger Penrose in the 1960s, that the laws of physics do not allow the formation of naked singularities.

cosmic horizon The outer edge or limit of the observable universe.

cosmic strings Very long, very dense (hypothetical) filaments that preserve inside themselves the mass density of the epoch of grand unification and the false vacuum of inflation. The main components of a time machine proposed in 1991 by J. Richard Gott.

cosmological principle Formulated in 1960 by Hermann Bondi, the assertion that we enjoy no privileged position in space, and that any conjectures about the nature of the universe should begin with this presumption. In 1968, Bondi and Thomas Gold broadened this tenet to create the "perfect cosmological principle," which asserted that there *is* no privileged place in space—also that there is no privileged moment in *time*, and that any conjectures about the nature of the universe should begin with these presumptions. Compare *steady-state theory*.

cosmology The science concerned with discovering and explaining the structure, composition, and past and future of the universe as a whole.

dark matter Matter that is detected only by its gravitational pull on visible matter (whose composition is unknown), yet comprises at least 90 percent of the matter in the universe.

determinism The doctrine that all events are, in principle, predictable as outcomes of causes that preceded them.

dimension A geometric axis in space or time.

Doppler shift Also termed *Doppler effect*. The change in the received frequency and wavelength of a wave that occurs when either the source or the receiver, or both, are in motion. Recession causes a shift toward longer wavelengths and lower frequencies (a redshift). Approach causes a shift toward shorter wavelengths and higher frequencies (a blueshift).

double-slit experiment The test, first performed by English physician and physicist Thomas Young in 1801, that demonstrates the wave nature of light. More provocatively, the test that involves a system with a part from which one extracts information and, in so doing, seems to determine the behavior of another part, although the distance between the parts is too great for a signal (even a signal traveling at the speed of light) to have traversed.

Einstein's field equations The fundamental formulas, developed by Albert Einstein and first published in 1915, for describing the gravitational effects produced by a given mass.

electromagnetic waves Waves of electrical and magnetic forces that include (ordered here from shorter to longer wavelengths) gamma rays, X-rays, ultraviolet radiation, visible light, infrared radiation, microwaves, and radio waves.

electron A light elementary particle with a negative electrical charge.

embedding diagram A type of illustration in which three-dimensional space is represented by a two-dimensional elastic surface, especially useful in depicting the effects of general relativity.

energy The capacity to perform work.

entropy The increase in the disorder of any closed system. More formally, a measurement of randomness in collections of particles, equal to the logarithm of the number of ways that the particles might be arranged without changing their collective, macroscopic appearance.

equivalence principle The tenet that the effects of acceleration are identical to the effects of a uniform gravitational field.

Euclidean geometry Flat geometry based upon the axioms of Euclid.

event A point in spacetime.

event horizon A black hole's "surface," the outermost edge of the roughly spherical region from which nothing—not even light—can escape. Also termed *absolute horizon* or *Schwarzschild horizon*.

Everett interpretation or **Everett-DeWitt interpretation** See *"many universes" interpretation*.

exotic matter Also termed *negative energy*. Material with a negative average energy density. In 1988, Mike Morris and Kip Thorne suggested that exotic matter might be used to hold open a wormhole mouth.

false vacuum A vacuum that has a negative pressure and generates a repulsive gravitational field.

flat universe A universe that is perfectly balanced between closed and open, such that its geometry is exactly Euclidean and triangles contain exactly 180 degrees. A flat universe would expand forever, but its rate of expansion would slow, gradually approaching zero but never reaching it.

free will The power to take an action that is intentional and uninhibited.

frequency The rate at which a wave oscillates. More precisely, the number of crests of a propagating wave that cross a given point in a given period of time.

general relativity See *relativity, general*.

geodesic A straight line in a curved space or spacetime.

Gödel's rotating universe Historically, the second solution to the Einstein field equations to describe a time machine. A model for a universe whose rotation tilts light cones forward in the direction of its rotation and so allows closed timelike curves everywhere. Thought to be unphysical. Compare *Van Stockum cylinder* and *Tipler cylinder*.

grandfather paradoxes Any of a variety of conditions that, by way of past-ward time travel, become self-contradictory: if I could travel back in time before my birth and kill my grandfather, I cannot have been born; if I was never born, then I cannot have lived to travel back in time to kill my grand-father; and so on. One of two large classes of time travel paradoxes. Compare *bootstrap paradoxes*.

grand unified theories Abbreviated GUTs. A set of theories developed in the mid 1970s that attempt to link three of the fundamental forces—electro-magnetic forces, strong forces, and weak forces (omitting only gravity)—and that purport to describe conditions in the very early universe.

gravitational wave A propagating ripple of spacetime curvature that travels at the speed of light.

GUTs See *grand unified theories*.

Hawking radiation The radiation emitted by black holes.

Heisenberg indeterminacy principle Also termed the *Heisenberg uncertainty principle*. The precept, associated with quantum mechanics and put forth by Werner Heisenberg, asserting that a given particle has no exact loca-tion or precise trajectory. Often described incorrectly as asserting merely that the position and trajectory of a particle cannot both be *known* with precision.

hyperspace A flat space that is utterly fictitious, but useful for imagining a region in which our universe's curved space is embedded—for example, the space through which a wormhole throat is imagined to be threaded.

hypothesis A scientific proposition that is supported by observational evi-dence and purports to explain a given phenomenon or set of phenomena. A hypothesis is neither as comprehensive nor as well established as a the-ory, although a set of related hypotheses may, over time, come to compose a theory.

indeterminacy principle See *Heisenberg indeterminacy principle*.

inertia Resistance to acceleration.

inertial reference frame A reference frame that does not rotate, that is unaf-fected by external forces, and whose motion is solely a product of its own inertia.

inflation The process that, by a repulsive gravitational field, may have driven the universe's early expansion.

inflationary theories The set of theories that the very early universe expanded far more rapidly than it does at present—at an exponential rather than a linear rate.

initial singularity See *singularity.*

jinn (sing. and pl.; also **jinni**) The hypothetical product of a bootstrap paradox; or, a hypothetical entity that seems to have been created from nothing
and whose worldline traces a closed timelike curve. Andrei Lossev and Igor
Novikov termed a material version of such an entity a "jinn of the first
kind," and a nonmaterial version a "jinn of the second kind."

Kelvin The temperature scale used by astronomers and astrophysicists. Zero
degrees Kelvin is absolute zero, the (unreachable) temperature at which
the thermal motion of atoms would cease altogether. The Heisenberg indeterminacy principle ensures that there will be some motion even at zero
degrees Kelvin. To convert Celsius to Kelvin, simply subtract 273.

Krasnikov tube A shrinkage of space in a region along a trajectory that would
allow faster travel in one direction between the points it connected than
would be possible otherwise. In principle, two Krasnikov tubes might be
used as a time machine.

laws of physics Fundamental principles by which one may deduce the behavior of the universe.

light Electromagnetic radiation with wavelengths that are detectable or
nearly detectable by the human eye.

light cone The surface representing all possible paths of light that could arrive
at or depart from a particular event. Arriving paths of material particles
would lie within the event's "past" light cone; departing paths of material
particles would lie within its "future" light cone. The paths of light rays
(both arriving and departing) would lie *on* the light cone.

lightlike interval See *null interval.*

light-year The distance that light travels in a year; 9,460,800,000,000 kilometers.

Lorentz transformations A set of equations that describe how the properties of
a moving object transform when viewed by observers moving at different
relative speeds.

"many universes" interpretation The interpretation of quantum mechanics
that asserts that each act of measurement causes the universe to split, or
after David Deutsch, that when a quantum alternative presents itself, universes partition themselves into groups in which each alternative is realized. Also termed the *Everett interpretation* or *Everett-DeWitt interpretation.*

mass The measure of the inertia of an object. Also the property that, according to Newtonian mechanics, produces and responds to a gravitational
field. Compare *stress-energy tensor.*

mass shell A spherical envelope of dense matter within which one would
experience time dilation without suffering tidal forces.

Minkowski diagram A depiction of spacetime using one or two dimensions of space as the x axis and time as the y axis. The other spatial dimensions are suppressed.

Morris-Thorne traversable wormholes A solution to the Einstein field equations found by Morris and Thorne in 1988 that described a wormhole whose parameters were set to allow passage and to ensure human "physiological comfort."

multiverse The (hypothetical) set of all universes that results from one of several scenarios: the "many universes" interpretation of quantum mechanics put forth in the 1950s by Hugh Everett III and later developed by Bryce DeWitt; the "eternal" inflationary phase posited by Andrei Linde and Alex Vilenkin in which universes are sprung from separate big bangs into different regions of spacetime; the result of the processes imagined by Alan Guth and Lee Smolin in which universes grow inside black holes; and the suggestion of Lisa Randall and Raman Sundrum that other universes are separated from us by another spatial dimension.

naked singularity A singularity existing outside an event horizon. See also *cosmic censorship conjecture*.

negative energy See *exotic matter*.

neutron An electrically neutral, massive particle found in the nuclei of atoms.

neutron star A star whose intense gravity has collapsed most of its matter into neutrons. A neutron star that is spinning rapidly and emitting radio waves is termed a "pulsar."

Newtonian mechanics The laws of physics based upon Newton's conception of space and time as absolute.

null interval Also termed *lightlike interval*. The interval connecting two points in spacetime between which a signal could travel at the speed of light.

observable universe The part of the universe that lies inside the cosmic horizon.

open universe A universe in which gravity is not strong enough to reverse its expansion and that thus continues to expand forever. Compare *closed universe*.

oscillating universe A cosmological model in which the universe's expansion eventually ceases, and the universe collapses and then rebounds in a new expansion phase as a new universe.

paradox A statement or proposition that contradicts itself; the source of much pleasure and disquiet among those considering pastward time travel. See also *bootstrap paradoxes* and *grandfather paradoxes*.

perfect cosmological principle See *cosmological principle*.

physics The scientific study of the interactions of matter, energy, space, and time.

Planck density 5×10^{93} grams per cubic centimeter.

Planck epoch The first 10^{-43} second of the existence of the universe, when the gravitational force acted as strongly as the other fundamental forces.

Planck length The length scale below which space as we know it ceases to exist. It is equal to 1.6×10^{-35} meters. Compare *Planck-Wheeler time*.

Planck-Wheeler time The time scale below which time as we know it ceases to exist. It is equal to $1/c$ times the Planck length, or about 10^{-43} second. Compare *Planck length*.

principle of equivalence See *equivalence principle*.

quantum foam Also termed *spacetime foam*. The theoretical, probabilistic structure of space that may compose the cores of singularities and that may comprise ordinary space on scales smaller than the Planck length.

quantum gravity Also termed the *theory of everything*. The (as yet nonexistent) theory that would eliminate disparities between relativity and quantum mechanics.

quantum mechanics Also termed *quantum physics*. The laws of physics that explain the behavior of the universe on very small scales (the scales of molecules, atoms, and electrons) and underlie the universe on larger scales. The laws that account in some way for vacuum fluctuations, the wave–particle duality, and various phenomena described by the Heisenberg indeterminacy principle. Compare *classical physics*.

quantum physics See *quantum mechanics*.

redshift The stretching of light waves traveling through space that occurs because empty space expands. This stretching produces a shift toward the longer wavelengths at the red end of the spectrum. Redshift of galaxies is taken as evidence that the universe is expanding.

reference frame An imaginary framework in which bodies may be said to exist and with which they move.

relativist A theoretical physicist whose work is based in general relativity.

relativity, general Einstein's theory of gravitation, put forth in 1915, which describes gravity not as an ordinary force, but rather as the warping (bending, twisting, or stretching) of spacetime.

relativity, special Einstein's theory of the electrodynamics of moving systems, put forth in 1905, which postulates that, because the speed of light is measured as the same value by all observers regardless of their speed or direction, measurements of time and distance are necessarily relative, depending upon the motion of the observer and the observed, and the laws of physics appear the same to all observers.

Roman ring First described by Matt Visser in 1995, a set of traversable wormholes arranged in a loop such that no single subset of wormholes is near to violating causality, yet the entire set allows pastward time travel because the back reaction of vacuum polarization, widely considered to enforce chronology protection, is kept as small as one might wish.

Schwarzschild horizon See *event horizon.*

second law of thermodynamics The law stating that entropy in a closed system can never decrease and is likely to increase.

self-consistency conjecture The speculation that pastward time travel need not violate causality, discussed in various forms by philosophers and authors of science fiction since at least the early twentieth century, and first framed in the language of physics in the 1980s by Igor Novikov.

semiclassical physics The set of laws that regard spacetime according to Newton and Einstein (that is, as well defined and classical) but that regard matter according to quantum mechanical laws (that is, as subject to uncertainty and quantum fluctuations).

semiclassical quantum gravity The approximation to the theory of quantum gravity that treats matter fields according to quantum mechanical laws and gravitational fields according to classical laws.

singularity A point of infinite curvature of space. Singularities appear at the centers of black holes and the very last moment of a collapsing universe. Using classical physics, in the 1960s and 1970s Roger Penrose and Stephen Hawking extrapolated back to "time zero" and arrived at a point of infinite density, infinite pressure, and infinite temperature, called the "initial singularity." The infinities in their result suggest that it is untrustworthy, and that classical physics is inadequate to the task of describing the phenomenon. See also *naked singularity.*

space Commonly, the three-dimensional arena of experience; more generally, such an arena defined by any number of dimensions.

spacelike interval An interval connecting two points in spacetime between which no signal can travel.

spacetime The four-dimensional arena necessary to depict events in special and general relativity.

spacetime foam See *quantum foam.*

spacetime interval The distance between two events in spacetime.

special relativity See *relativity, special.*

steady-state theory The theory of cosmology—developed in the late 1940s by Hermann Bondi, Thomas Gold, and Fred Hoyle—asserting that the universe appears the same at all moments in its history and that new matter is constantly being created, filling the expanding spaces between the galaxies. Compare *cosmological principle*.

stress-energy tensor In general relativity, the source of the gravitational field. (In Newtonian mechanics, mass is the source of gravity.)

string theory A proposal for ultimate natural laws or a "theory of everything." Its fundamental entity is string with a length of 10^{-33} centimeter and zero thickness.

subluminary Slower than the speed of light. Compare *supraluminary*.

sum-over-histories interpretation A probabilistic interpretation of a system's past in which the history is reconstructed in terms of each possible path and its relative likelihood.

supraluminary Faster than the speed of light. Compare *subluminary*.

tesseract The extension of a cube into four dimensions.

theoretical physics A field of study that takes results of experiments conducted by experimental physicists, relates them to other results, and explains them in broad contexts. One of two broad domains of physics (the other being experimental physics).

theory A set of hypotheses that are supported by observational evidence and purport to explain a phenomenon or set of phenomena. A theory can be tested and may thereby gain legitimacy over time. Strictly speaking, a theory can be proven false but can never be proven true.

theory of everything See *quantum gravity*.

time The dimension that distinguishes past, present, and future.

time dilation The slowing of the passage of time achieved by linear motion outbound and inbound at relativistic velocities, circular motion at relativistic velocities, and/or subjection to a strong gravitational field like that in the vicinity of a black hole or inside a mass shell.

timelike interval The interval separating an event and a point inside its light cone. An interval connecting two points in spacetime between which a message could travel at less than the speed of light.

Tipler cylinder Probably the third solution to Einstein's field equations to describe a time machine. An infinitely long massive cylinder rotating so rapidly that its frame dragging would warp a closed timelike curve completely around it. Thought to be unphysical. Compare *Van Stockum cylinder* and *Gödel's rotating universe*.

uncertainty principle See *Heisenberg indeterminacy principle*.

universe Everything that exists; more technically, the set of all observable and potentially observable phenomena.

vacuum In Newtonian physics, a space devoid of matter.

vacuum fluctuation An unpredictable event occurring as a result of the probabilistic nature of the quantum world. Particles may materialize along with their antiparticles, and the values of fields may fluctuate.

Van Stockum cylinder The first solution to Einstein's field equations to describe a time machine. An infinitely long cylinder of dust rotating so rapidly that its frame dragging would warp a closed null curve completely around it, and a signal traveling along its "surface" would return to its sender from the direction opposite to that from which it was sent. Thought to be unphysical. Compare *Gödel's rotating universe* and *Tipler cylinder*.

wave–particle duality The observation that quanta exhibit characteristics of both waves and particles. See also *double-slit experiment*.

weak energy condition The "rule" requiring that, for all observers, the local energy in all spacetime locations must be greater than or equal to zero.

worldline The path traced by an object in spacetime.

wormhole A connection between two places that, in three-dimensional space, are disparate.

BIBLIOGRAPHY

Abbott, Edwin A. *The Annotated Flatland: A Romance of Many Dimensions.* Introduction and notes by Ian Stewart. Cambridge, MA: Perseus Publishing, 2002.

Abell, George O. *Exploration of the Universe.* 3rd ed. New York: Holt, Rinehart and Winston, 1975.

Adams, Fred, and Greg Laughlin. *The Five Ages of the Universe: Inside the Physics of Eternity.* New York: Free Press, 1999.

Alcubierre, Miguel. "The Warp Drive: Hyper-fast Travel within General Relativity." *Classical and Quantum Gravity* 11 (1994): L73–L77.

Allen, B., and J. Simon. "Time Travel on a String." *Nature* 357 (May 1992): 19–21.

Allman, William F. "Playing Fast and Loose with Time." *U.S. News & World Report*, December 19, 1988, 62.

Anstey, F. [Thomas Anstey Guthrie]. *Tourmalin's Time Cheques.* New York: D. Appleton and Company, 1891.

Augustine. *Confessions.* Translated by R. S. Pine-Coffin. London: Penguin Books, 1961.

Baeyer, Hans Christian von. *The Emergence of the Visible Microworld.* New York: Random House, 1992.

Ball, John A. "Extraterrestrial Intelligence: Where Is Everybody?" *Icarus* 19 (1973): 347.

Barrow, Isaac. *The Geometrical Lectures of Isaac Barrow.* Translated, with notes and proofs, and a discussion on the advance made therein of the work of his predecessors in the infinitesimal calculus, by J. M. Child. Chicago: Open Court Publishing Company, 1916.

Barrow, John D. *The Constants of Nature: From Alpha to Omega—The Numbers That Encode the Deepest Secrets of the Universe.* New York: Pantheon, 2002.

Barrow, J. D., and Frank J. Tipler. *The Anthropic Cosmological Principle.* Oxford: Clarendon Press, 1986.

Bergson, Henri. *Time and Free Will; an Essay on the Immediate Data of Consciousness.* Translated by F. L. Pogson. 1910. Reprint, London: G. Allen & Company, 1959.

Bernal, J. D. *The World, the Flesh and the Devil: An Enquiry into the Future of the Three Enemies of the Rational Soul.* London: K. Paul, Trench, Trubner & Company, 1929.

Borges, Jorge Luis. *Ficciones.* Edited and with an introduction by Anthony Kerrigan. New York: Grove Press, 1962.

Born, Max. *The Born-Einstein Letters.* New York: Walker, 1971.

Boslough, John. *Masters of Time: Cosmology at the End of Innocence.* Reading, MA: Addison-Wesley, 1992.

Boulware, David G. "Quantum Field Theory in Spaces with Closed Timelike Curves." *Physical Review D* 46, no. 10 (November 15, 1992): 4421–4442.

Boyer, Carl B. *A History of Mathematics.* 2nd ed. Revised by Uta C. Merzbach. John Wiley & Sons, 1989.

Browne, Malcolm W. "3 Scientists Say Travel in Time Isn't So Far Out." *New York Times*, November 22, 1988, sec. C.

Carlini, A., V. P. Frolov, M. B. Mensky, I. D. Novikov, and H. H. Soleng. "Time Machines: The Principle of Self-Consistency as a Consequence of the Principle of Minimal Action." *International Journal of Modern Physics D* 4, no. 5 (1995): 557–580.

Carlini, A., and I. D. Novikov. "Time Machines and the Principle of Self-Consistency as a Consequence of the Principle of Stationary Action." *International Journal of Modern Physics D* 5, no. 5 (1996): 445–479.

Carroll, Lewis. *The Complete Sylvie and Bruno.* San Francisco: Mercury House, 1991.

Carroll, Sean M., Edward Farhi, and Alan H. Guth. "An Obstacle to Building a Time Machine." *Physical Review Letters* 68, no. 3 (January 20, 1992): 267–269.

Carter, Brandon. "Complete Analytic Extension of the Symmetry Axis of Kerr's Solution of Einstein's Equations." *Physical Review* 141 (1966): 1242–1247.

———. "Global Structure of the Kerr Family of Gravitational Fields." *Physical Review* 174 (1968): 1559–1571.

————. "Large Number Coincidences and the Anthropic Principle." In *Confrontations of Cosmological Theories with Observational Data*, edited by M. S. Longair, 291–298. Boston: Reidel, 1974.

Casati, R., and A. C. Varzi. "Time Reversal and the K^0 Meson Decays. II." *Physical Review Letters* 22, no. 11 (March 17, 1969): 554–556.

Cassidy, Michael J., and Stephen W. Hawking. "Models for Chronology Selection." *Physical Review D* 57, no. 4 (February 15, 1998): 2372–2380.

Chandrasekhar, Subrahmanyan. *Eddington: The Most Distinguished Astrophysicist of His Time.* London: Cambridge University Press, 1983.

————. *The Mathematical Theory of Black Holes.* Oxford: Clarendon Press, 1983.

Clarke, Arthur C. *The Collected Stories of Arthur C. Clarke.* New York: Tom Doherty Associates, 2000.

————. *Profiles of the Future: An Inquiry into the Limits of the Possible.* New York: Holt, Rinehart and Winston, 1984.

Cooper, Henry S. F. *The Search for Life on Mars.* New York: Holt, Rinehart and Winston, 1980.

Cramer, John G., Robert L. Forward, Michael S. Morris, Matt Visser, Gregory Benford, and Geoffrey A. Landis. "Natural Wormholes as Gravitational Lenses." *Physical Review D* 51, no. 6 (March 15, 1995): 3117–3120.

Cutler, Curt. "Global Structure of Gott's Two-String Spacetime." *Physical Review D* 45, no. 2 (January 1992): 487–494.

Davidson, Keay. *Carl Sagan: A Life.* New York: John Wiley & Sons, 1999.

Davies, P. C. W., and J. R. Brown, eds. *The Ghost in the Atom: A Discussion of the Mysteries of Quantum Physics.* Cambridge: Cambridge University Press, 2002.

Davies, Paul. *About Time: Einstein's Unfinished Revolution.* New York: Simon & Schuster, 1995.

————. *The Last Three Minutes: Conjectures about the Ultimate Fate of the Universe.* New York: HarperCollins, 1997.

————. *Other Worlds: Space, Superspace, and the Quantum Universe.* Reprint ed. New York: Penguin Books, 1997.

Dawson, John W., Jr. *Logical Dilemmas: The Life and Work of Kurt Gödel.* Wellesley, MA: A. K. Peters, 1997.

Deser, S., R. Jackiw, and G. 't Hooft. "Physical Cosmic Strings Do Not Generate Closed Timelike Curves." *Physical Review Letters* 68, no. 3 (January 20, 1992): 263–266.

Deutsch, David. *The Fabric of Reality: The Science of Parallel Universes—and Its Implications.* New York: Allen Lane, 1997.

————. "Quantum Mechanics Near Closed Timelike Lines." *Physical Review D* 44, no. 10 (November 15, 1991): 3197–3217.

Deutsch, David, and Michael Lockwood. "The Quantum Physics of Time Travel." *Scientific American*, March 1994, 50–58.

DeWitt, Bryce S., and Neil Graham, eds. *The Many-Worlds Interpretation of Quantum Mechanics*. Princeton, NJ: Princeton University Press, 1973.

Dick, Steven J. *The Biological Universe: The Twentieth-Century Extraterrestrial Life Debate and the Limits of Science*. New York: Cambridge University Press, 1996.

————. *Plurality of Worlds: The Origins of the Extraterrestrial Life Debate from Democritus to Kant*. New York: Cambridge University Press, 1982.

Dyson, Freeman J. *Disturbing the Universe*. New York: Harper & Row, 1979.

————. "Search for Artificial Stellar Sources of Infrared Radiation." *Science* 131 (June 3, 1960): 1667–1668.

————. "Time without End: Physics and Biology in an Open Universe." *Reviews of Modern Physics* 51, no. 3 (July 1979): 447–460.

Dyson, Lisa. "Chronology Protection in String Theory." *Journal of High Energy Physics*, March 2004.

Echeverria, F., G. Klinkhammer, and K. Thorne. "Billiard Balls in Wormhole Spacetimes with Closed Timelike Curves: Classical Theory." *Physical Review D* 44, no. 4 (August 15, 1991): 1077–1099.

Eddington, Arthur. "The End of the World from the Standpoint of Mathematical Physics." *Nature* 127 (1931): 447–453.

————. *The Nature of the Physical World*. New York: Macmillan Company, 1928.

Einstein, Albert. *The Expanded Quotable Einstein*. Collected and edited by Alice Calaprice. Princeton, NJ: Princeton University Press, 2000.

————. "Die Grundlage der allgemeinen Relativitätstheorie." *Annalen der Physik* 49 (1916).

————. "Physics and Reality." *Journal of the Franklin Institute* 221, no. 3 (March 1936): 349–382.

————. *Relativity: The Special and the General Theory*. New York: Crown Publishers, 1961.

————. "Religion and Science." *New York Times Magazine*, November 9, 1930, 1–4.

————. *Sidelights on Relativity*. New York: E. P. Dutton and Co., 1923.

Einstein, A. [Albert], B. Podolsky, and N. Rosen. "Can Quantum-Mechanical Description of Physical Reality Be Considered Complete?" *Physical Review* 47 (1935): 777.

Einstein, Albert, and Nathan Rosen. "The Particle Problem in the General Theory of Relativity." *Physical Review* 48 (1935): 73–77.

Eiseley, Loren C. *The Night Country.* New York: Scribner, 1971.

Escher, M. C. *Escher on Escher: Exploring the Infinite.* Translated by Karin Ford. New York: Harry N. Abrams Publishers, 1989.

Everett, Allen E. "Time Travel Paradoxes, Path Integrals and the Many Worlds Interpretation of Quantum Mechanics." *Physical Review D* 69, no. 12 (June 15, 2004): 124023-1–124023-14.

———. "Warp Drive and Causality." *Physical Review D* 53, no. 12 (June 15, 1996): 7365–7368.

Everett, Allen E., and Thomas A. Roman. "Superluminal Subway: The Krasnikov Tube." *Physical Review D* 56, no. 4 (August 15, 1997): 2100–2108.

Everett, Hugh, III. "'Relative State' Formulation of Quantum Mechanics." *Reviews of Modern Physics* 29, no. 3 (July 1957): 454–462.

Farhi, Edward H., and Alan H. Guth. "An Obstacle to Creating a Universe in the Laboratory." *Physics Letters* 183B (1987): 149.

Ferris, Timothy. *The Mind's Sky: Human Intelligence in a Cosmic Context.* New York: Bantam Books, 1992.

———. *The Whole Shebang: A State-of-the-Universe(s) Report.* New York: Simon & Shuster, 1997.

Feynman, Richard P., and A. R. Hibbs. *Quantum Mechanics and Path Integrals.* New York: McGraw-Hill, 1965.

Feynman, Richard P., Robert B. Leighton, and Matthew Sands. *The Feynman Lectures on Physics*, Vol. 1. Reading, MA: Addison-Wesley, 1963.

———. *The Feynman Lectures on Physics*, Vol. 2. Reading, MA: Addison-Wesley, 1963.

Flamm, Ludwig. "Beitrage zur einsteinschen Gravitationstheorie." *Physik Zeitschrift* 17 (1916): 448–454.

Flynn, John L. "Time Travel Literature," www.towson.edu/~flynn/timetv.html (accessed September 4, 2006).

Folger, Tim. "Physics' Best Kept Secret." *Discover*, September 2001.

Ford, Lawrence A., and Thomas A. Roman. "Negative Energy, Wormholes and Warp Drive." *Scientific American*, January 2000, 46–53.

———. "Quantum Field Theory Constrains Traversable Wormhole Geometries." *Physical Review D* 53, no. 10 (May 15, 1996): 5496–5507.

———. "The Quantum Interest Conjecture." *Physical Review D* 60, no. 10 (November 15, 1999): 104018-1–104018-8.

Forward, Robert L. *Indistinguishable from Magic: Speculations and Visions of the Future.* Riverdale, NY: Baen Publishing Enterprises, 1995.

———. "When You Live upon a Star." *New Scientist*, December 24/31, 1987: 36–38.

Friedman, John L. "Back to the Future." *Nature* 336 (November 1988): 305–306.

Friedman, John L., Michael S. Morris, Igor D. Novikov, Fernando Echeverria, Gunnar Klinkhammer, Kip S. Thorne, and Ulvi Yurtsever. "Cauchy Problem in Spacetimes with Closed Timelike Curves." *Physical Review D* 42, no. 6 (September 15, 1990): 1915–1930.

Friedman, John L., Nicolas J. Papastamatiou, and Jonathan Z. Simon. "Failure of Unitarity for Interacting Fields on Spacetimes with Closed Timelike Curves." *Physical Review D* 46, no. 10 (November 15, 1992): 4456–4469.

Frolov, Valery P., and Igor D. Novikov. "Physical Effects in Wormholes and Time Machines." *Physical Review D* 42, no. 4 (August 15, 1990): 1057–1065.

Fuller, Robert W., and John A. Wheeler. "Causality and Multiply-Connected Space-Time." *Physical Review* 128, no. 2 (October 1962): 919–929.

Gardner, Martin. "Mathematical Games: On Altering the Past, Delaying the Future and Other Ways of Tampering with Time." *Scientific American*, March 1979, 21–30.

———. *The New Ambidextrous Universe: Symmetry and Assymmetry from Mirror Reflections to Superstrings*. 3rd rev. ed. New York: W. H. Freeman and Company, 1990.

Gauss, Carl Friedrich. *General Investigations of Curved Surfaces*. Edited with an Introduction and Notes by Peter Pesic; translated by Adam Hiltebeitel and James Morehead. Mineola, NY: Dover, 2005.

Geroch, Robert. "Topology in General Relativity." *Journal of Mathematical Physics*, 8 (April 1967): 782–786.

Gilster, Paul. *Centauri Dreams: Imagining and Planning Interstellar Exploration*. New York: Copernicus Books, 2004.

Gleick, James. *Isaac Newton*. New York: Pantheon, 2003.

Gödel, Kurt. "An Example of a New Type of Cosmological Solution of Einstein's Field Equations of Gravitation. *Reviews of Modern Physics* 21, no. 3 (July 1949): 447–450.

———. "Lecture on Rotating Universes" (1949). In *Collected Works*, Vol. 3, *Unpublished Essays and Lectures*. Edited by Solomon Feferman, John W. Dawson, Jr., Warren Goldfarb, Charles Parsons, and Robert M. Soloway. New York: Oxford University Press, 1995.

———. "A Remark about the Relationship between Relativity Theory and Idealistic Philosophy." In *Albert Einstein: Philosopher-Scientist*, edited by Paul

Arthur Schilpp, 687–688. Evanston, Illinois: Library of Living Philosophers, 1949.

Gold, Thomas. "Mother and Baby Paradox." *Nature* 256 (July 10, 1975): 113.

———. "Rotating Neutron Stars as Origin of Pulsating Radio Sources." *Nature* 218 (May 25, 1968): 731.

González-Díaz, Pedro F. "Ringholes and Closed Timelike Curves." *Physical Review D* 54, no. 10 (November 15, 1996): 6122–6131.

Gott, J. Richard. "Closed Timelike Curves Produced by Pairs of Moving Cosmic Strings: Exact Solutions." *Physical Review Letters* 66, no. 9 (March 4, 1991): 1126–1129.

———. "Implications of the Copernican Principle for Our Future Prospects." *Nature* 363 (May 27, 1993): 315–319.

———. *Time Travel in Einstein's Universe.* Boston: Houghton Mifflin Company, 2001.

Gott, J. Richard, and Li-Xin Li. "Can the Universe Create Itself?" *Physical Review D* 58, no. 2 (July 15, 1998).

Grant, James D. E. "Cosmic Strings and Chronology Protection." *Physical Review D* 47, no. 6 (March 15, 1993): 2388–2394.

Gray, Jeremy. *Ideas of Space: Euclidean, Non-Euclidean and Relativistic.* Oxford: Oxford University Press, 1990.

Gribbin, John. *In Search of the Edge of Time: Black Holes, White Holes, Wormholes.* London: Penguin Books, 1998.

———. *Schrödinger's Kittens and the Search for Reality: Solving the Quantum Mysteries.* New York: Little, Brown, 1995.

Grosskurth, Phyllis. *Havelock Ellis: A Biography.* New York: Alfred A. Knopf, 1980.

Guth, Alan H. *The Inflationary Universe: The Quest for a New Theory of Cosmic Origins.* Reading, MA: Addison-Wesley, 1997.

Haldane, J. B. S. [John Burdon Sanderson]. *Daedalus: Or, Science and the Future; A Paper Read to the Heretics, Cambridge on February 4th, 1923.* New York: E. P. Dutton, 1923.

Hall, Stephen S. "Professor Thorne's Time Machine." *California*, October 1989, 68–75 and 158–162.

Harrison, Edward. *Cosmology: The Science of the Universe.* 2nd ed. Cambridge: Cambridge University Press, 2000.

———. "The Natural Selection of Universes Containing Intelligent Life." *Quarterly Journal of the Royal Astronomical Society* 36, no. 3 (1995): 193.

Hart, Michael H., and Ben Zuckerman, eds. *Extraterrestrials—Where Are They?* New York: Pergamon Press, 1982.

Hawking, Stephen. *Black Holes and Baby Universes and Other Essays*. New York: Bantam Books, 1993.

———. *A Brief History of Time*. Updated and expanded tenth anninversary ed. New York: Bantam Books, 1998.

———. "The Chronology Protection Conjecture." In *Sixth Marcel Grossman Meeting, Proceedings, Kyoto, Japan, 1991*, edited by H. Sato. Singapore: World Scientific, 1992.

———. "The Chronology Protection Conjecture." *Physical Review D* 46, no. 2 (July 15, 1992): 603–611.

Hawking, Stephen W., and G. F. R. Ellis. *The Large Scale Structure of Space-Time*. Cambridge: Cambridge University Press, 1973.

Hawking, Stephen W., Kip S. Thorne, Igor Novikov, Timothy Ferris, and Alan Lightman. Introduction by Richard Price. *The Future of Spacetime*. New York: W. W. Norton & Company, 2002.

Heinlein, Robert. *The Fantasies of Robert A. Heinlein*. New York: Tor, 1999.

Hinton, C. H. "A Mechanical Pitcher." *Harper's Weekly*, March 1897, 301–302.

———. *Scientific Romances*. London: Swan Sonnenschein & Co., 1886.

———. *Speculations on the Fourth Dimension: Selected Writings*. Edited by R. v. B. Rucker. New York: Dover, 1980.

———. "What Is the Fourth Dimension?" [From *Scientific Romances*, Vol. 1 (1884).] In *Speculations on the Fourth Dimension: Selected Writings of Charles H. Hinton*, edited by Rudolf v. B. Rucker, pp. 1–22. New York: Dover, 1980, http://www.ibiblio.org/eldritch/chh/h1.html (accessed August 20, 2006).

Hiscock, William A., and D. A. Konkowski. "Quantum Vacuum Energy in Taub-NUT (Newman-Unti-Tamborino)-Type Cosmologies." *Physical Review D* 26, no. 6 (September 15, 1982): 1225–1230.

Hofstadter, Douglas R. *Gödel, Escher, Bach: An Eternal Golden Braid*. New York: Basic Books, 1979.

Hollingdale, Stuart. *Makers of Mathematics*. London: Penguin Books, 1989.

Hoyle, Fred. "The Universe: Past and Present Reflections." *Engineering and Science*, November 1981.

Irion, Robert. "Kip Thorne: The Shaman of Space and Time." *Science* 290 (November 24, 2000): 1488–1491.

Islam, Jamal N. "Possible Ultimate Fate of the Universe." *Quarterly Journal of the Royal Astronomical Society* 18 (March 1977): 3–8.

Jammer, Max. *Einstein and Religion*. Princeton, NJ: Princeton University Press, 1999.

———. *The Philosophy of Quantum Mechanics; the Interpretations of Quantum Mechanics in Historical Perspective*. New York: Wiley, 1974.

Jeans, Sir James. *The Mysterious Universe.* New York: Macmillan Company, 1930.

Jemmin, Max. *Einstein and Religion: Physics and Theology.* Princeton, NJ: Princeton University Press, 1999.

Johnson, George. "The Odds of God." Review of *The Physics of Immortality,* by Frank J. Tippler. *New York Times Book Review* (October 9, 1994): 15–16.

Kaku, Michio. *Hyperspace: A Scientific Odyssey through Parallel Universes, Time Warps and the Tenth Dimension.* New York: Oxford University Press, 1994.

Kant, Immanuel. "Thoughts on the True Estimation of Living Forces." In *Kant's Inaugural Dissertation and Early Writings on Space,* translated by J. Handyside. Chicago: University of Chicago Press, 1929.

Kardashev, N. S. "Transmission of Information by Extraterrestrial Civilizations." *Soviet Astronomy* 8, no. 2 (September–October 1964): 217–221.

Kay, Bernard S., Marek J. Radzikowski, and Robert M. Wald. "Quantum Field Theory on Spacetimes with a Compactly Generated Cauchy Horizon." *Communications in Mathematical Physics* 183 (1997): 533–556.

Kidd, John, et al. *Bridgewater Treatises on the Power Wisdom and Goodness of God As Manifested in the Creation: On the Adaptation of External Nature to the Physical Condition of Man: Principally with Reference to the Supply of His Wants and the Exercise of His Intellectual Faculties.* London: H. G. Bohn, 1852.

Kim, Sung-Won, and Kip S. Thorne. "Do Vacuum Fluctuations Prevent the Creation of Closed Timelike Curves?" *Physical Review D* 43, no. 12 (June 15, 1991): 3929–3947.

Krasnikov, Sergei S. "Hyperfast Travel in General Relativity." *Physical Review D* 57, no. 8 (April 15, 1998): 4760–4766.

Krauss, Lawrence M., and Glenn D. Starkman. "The Fate of Life in the Universe." *Scientific American,* December 2002, 50–57.

———. "Life, the Universe, and Nothing: Life and Death in an Ever-Expanding Universe." *Astrophysical Journal* 531 (2000): 22–30.

Landau, Lev Davidovich. "Origin of Stellar Energy." *Nature* 141 (1938): 333–334.

Lemonick, Michael D. "How to Go Back in Time." *Time* (May 13, 1991): 74.

Leslie, John. *Universes.* New York: Routledge, 1989.

Lewis, David. "The Paradoxes of Time Travel." *American Philosophical Quarterly* 13, no. 2 (April 1976).

Li, Li-Xin. "Must Time Machines Be Unstable against Vacuum Fluctuations?" *Classical and Quantum Gravity* 13 (1995): 2563.

———. "New Light on Time Machines: Against the Chronology Protection Conjecture." *Physical Review D* 50, no. 10 (November 15, 1994): R6037–R6040.

————. "Time Machines Constructed from Anti–de Sitter Space." *Physical Review D* 59, no. 8 (April 15, 1999): 084016-1–084016-15.

Li, Li-Xin, Jian-Mei Xu, and Liao Liu. "Complex Geometry, Quantum Tunneling, and Time Machines." *Physical Review D* 48, no. 10 (November 15, 1993): 4735–4737.

Lightman, Alan P., William H. Press, Richard H. Price, and Saul. A. Teukolsky. *Problem Book on Relativity and Gravitation.* Princeton, NJ: Princeton University Press, 1975.

Lippincott, Kristen. *The Story of Time.* With Umberto Eco, E. H. Gombrich, and others. London: Merrell Holberton Publishers, 1999.

Liversidge, Anthony. "Interview: Frank Tipler." *Omni* 17, no. 1 (October 1994): 89.

Livio, Mario. *The Accelerating Universe: Infinite Expansion, the Cosmological Constant, and the Beauty of the Cosmos.* New York: John Wiley & Sons, 2000.

Locke, John. *An Essay concerning Human Understanding.* New York: Dover, 1959.

————. *Second Treatise of Government.* Edited with an introduction by C. B. Macpherson. Indianapolis, IN: Hackett, 1980.

Lorentz, H. A., A. Einstein, H. Minkowski, and H. Weyl. *The Principle of Relativity: A Collection of Original Memoirs on the Special and General Theory of Relativity.* Translated by W. Perrett and G. B. Jeffery. New York: Dover, 1923.

Lossev, Andrei, and Igor D. Novikov. "The Jinn of the Time Machine: Nontrivial Self-Consistent Solutions." *Classical and Quantum Gravity* 9 (October 1992): 2309–2321.

Lyutikov, Maxim. "Vacuum Polarization at the Chronology Horizon of Roman Spacetime." *Physical Review D* 49, no. 8 (April 15, 1994): 4041–4148.

Maxwell, James Clerk. *The Scientific Papers of James Clerk Maxwell,* Vol. 2. Edited by W. D. Niven. New York: Dover, 1890.

McNeil, Russel. "Wormholes in Space Turn Time Inside-Out." *Toronto Star* (December 4, 1988), sec. H.

"The Mechanical Baseball Pitcher." *Scientific American,* June 26, 1897, 409.

Meschkowski, Herbert. *Evolution of Mathematical Thought.* San Francisco: Holden-Day, 1964.

Michael, Donald N. "Proposed Studies on the Implications of Peaceful Space Activities for Human Affairs." Prepared for the [Committee on Long Range Studies of the] National Aeronautics and Space Administration by the Brookings Institution. Report of the Committee on Science and Astronautics, U.S. House of Representatives, Eighty-seventh Congress, first session. Washington, DC: U.S. Government Printing Office, 1961.

Minkowski, Hermann. *Gesammelte Abhandlungen von Hermann Minkowski*, Vol. 2. Leipzig: Teubner, 1911.

Misner, C. W., K. S. Thorne, and J. A. Wheeler. *Gravitation*. San Francisco: W. H. Freeman and Company, 1973.

Misner, Charles W., and John A. Wheeler. "Classical Physics as Geometry." *Annals of Physics*, 2, no. 6 (December 1957), 525–603.

Mitchell, Edward Page. "The Clock That Went Backwards." *Sun*, September 18, 1881.

Morris, Michael S., and Kip S. Thorne. "Wormholes in Spacetime and Their Use for Interstellar Travel: A Tool for Teaching General Relativity." *American Journal of Physics* 56 (May 1988): 395–412.

Morris, Michael S., Kip S. Thorne, and Ulvi Yurtsever. "Wormholes, Time Machines, and the Weak Energy Condition." *Physical Review Letters* 61, no. 13 (September 26, 1988): 1446–1449.

Nahin, Paul J. *Time Machines: Time Travel in Physics, Metaphysics, and Science Fiction*. Foreword by Kip S. Thorne. New York: Springer-Verlag, 1999.

Newcomb, Simon. "Modern Mathematical Thought." *Nature*, February 1, 1894.

Newton, Isaac. *Sir Isaac Newton's Mathematical Principles of Natural Philosophy and His System of the World*. Translated into English by Andrew Mottee in 1929. The translations revised, and supplied with an historical and explanatory index, by Florian Cajori. Berkeley: University of California Press, 1916.

North, John D. *The Measure of the Universe: A History of Modern Cosmology*. Oxford: Clarendon Press, 1965.

Novikov, Igor D. "An Analysis of the Operation of a Time Machine." *Soviet Journal of Experimental and Theoretical Physics* 68 (1989): 439–443.

———. *Evoliutsiia vselennoi*. Moscow: Nauka, 1979.

———. *Evolution of the Universe*. Translated by M. M. Basko. Cambridge: Cambridge University Press, 1983.

———. *The River of Time*. Cambridge: Cambridge University Press, 1998.

———. "Time Machine and Self-Consistent Evolution in Problems with Self-Interaction." *Physical Review D* 45, no. 6 (March 15, 1992): 1989–1994.

Oppenheimer, J. Robert, and Hartland Snyder. "On Continued Gravitational Contraction." *Physical Review* 56 (1939): 455.

Oppenheimer, J. Robert, and George Volkoff. "On Massive Neutron Cores." *Physical Review* 55 (1939): 374–381.

Ori, Amos. "A Class of Time-Machine Solutions with a Compact Vacuum Core." *Physical Review Letters* 95, no. 3 (July 7, 2005).

Overbye, Dennis. "In Aspen, Physics on a High Plane." *New York Times*, August 28, 2001, sec. F.

———. *Lonely Hearts of the Cosmos: The Scientific Quest for the Secret of the Universe.* New York: HarperCollins, 1991.

———. "Peering through the Gates of Time." *New York Times*, March 12, 2002.

Pagels, Heinz R. *The Cosmic Code: Quantum Mechanics as the Language of Nature.* New York: Bantam Books, 1983.

———. *Perfect Symmetry: The Search for the Beginning of Time.* New York: Bantam Books, 1985.

Pais, Abraham. *Einstein Lived Here.* Oxford: Clarendon Press, 1994.

———. *"Subtle Is the Lord—": The Science and the Life of Albert Einstein.* Oxford: Oxford University Press, 1982.

Paley, William. *Natural Theology: Evidences of the Existence and Attributes of the Deity Collected from the Appearances of Nature.* Oxford: J. Vincent, 1826.

Penrose, L. S., and Roger Penrose. "Impossible Objects, a Special Type of Visual Illusion." *British Journal of Psychology* 49, no. 1 (February 1958): 31–33.

Penrose, Roger. "Singularities and Time-Asymmetry." In *General Relativity: An Einstein Centenary Survey,* edited by S. W. Hawking and W. Israel, 531–638. New York: Cambridge University Press, 1979.

Peter Damian. *Letters 91–120.* Translated by Owen J. Blum. Washington, DC: Catholic University of America Press, 1998.

Pfenning, M. J., and Lawrence Ford. "The Unphysical Nature of Warp Drive." *Classical and Quantum Gravity* 14, no. 7 (July 1997): 1743–1751.

Price, Richard H., and Kip S. Thorne. "The Membrane Paradigm for Black Holes." *Scientific American*, April 1988: 69–77.

Rees, Martin J. *Our Cosmic Habitat.* Princeton, NJ : Princeton University Press, 2001.

———. *Our Final Hour: A Scientist's Warning: How Terror, Error, and Environmental Disaster Threaten Humankind's Future in This Century—On Earth and Beyond.* New York: Basic Books, 2003.

Reiganum, M. R. "Is Time Travel Possible? A Financial Proof." *Journal of Portfolio Management* 13 (1986): 10–12.

Remnant, Peter. "Peter Damian: Could God Change the Past?" *Canadian Journal of Philosophy*, 8, no. 2 (June 1978): 259–268.

Rigden, John S. *Einstein 1905: The Standard of Greatness.* Cambridge, MA: Harvard University Press, 2005.

Roman, Thomas. "Inflating Lorentzian Wormholes." *Physical Review D* 47, no. 4 (February 15, 1993): 1370–1379.

———. "On the 'Averaged Weak Energy Condition' and Penrose's Singularity Theorem." *Physical Review D* 37, no. 2 (January 15, 1988): 546–548.

———. "Quantum-Stress-Energy Tensors and the Weak Energy Condition." *Physical Review D* 33, no. 12 (June 15, 1986): 3526–3533.

———. "Some Thoughts on Energy Conditions and Wormholes" (September 23, 2004), http://arxiv.org/abs/gr-qc/0409090 (acessed March 25, 2006).

Rosenfeld, B. A. [Boris Abramovich]. *A History of Non-Euclidean Geometry: Evolution of the Concept of a Geometric Space.* Translated by Abe Shenitzer with the editorial assistance of Hardy Grant. New York: Springer Verlag, 1988.

Rucker, Rudolph v. B. *Geometry, Relativity and the Fourth Dimension.* New York: Dover, 1977.

———, ed. *Speculations on the Fourth Dimension: Selected Writings of Charles H. Hinton.* New York: Dover, 1980.

Russell, Bertrand. *Why I Am Not a Christian.* New York: Simon & Schuster, 1957.

Sagan, Carl. *The Cosmic Connection; an Extraterrestrial Perspective.* Produced by Jerome Agel. Garden City, NY: Anchor Press, 1973.

Sakharov, Andrei. *Memoirs.* New York: Alfred A. Knopf, 1990.

Schilling, Govert. "String Revival: Are Cosmic Strings behind Unusual Lensing Effects?" *Scientific American*, February 7, 2005, 25.

Schilpp, Paul Arthur, ed. *Albert Einstein: Philosopher-Scientist.* London: Cambridge University Press, 1970.

Semeniuk, Ivan. "No Going Back." *New Scientist*, September 20, 2003, 28–32.

Shikhovtsev, Eugene B., and Kenneth W. Ford. "Biographical Sketch of Hugh Everett, III" (2003), http://space.mit.edu/home/tegmark/everett (accessed July 15, 2006).

Smolin, Lee. *Three Roads to Quantum Gravity.* New York: Basic Books 2001.

Stapledon, Olaf. *Last and First Men & Star Maker.* New York: Dover, 1968.

Sushkov, Sergey V. "Chronology Protection and Quantized Fields: Complex Automorphic Scalar Field in Misner Space." *Classical and Quantum Gravity* 14, no. 2 (February 1997): 523–534.

Tanaka, Tsunefumi, and William A. Hiscock. "Massive Scalar Field in Multiply Connected Flat Spacetimes." *Physical Review D* 52, no. 8 (October 15, 1995): 4503–4511.

Thorne, Kip S. *Black Holes and Time Warps: Einstein's Outrageous Legacy.* New York: W. W. Norton & Company, 1994.

————. "Closed Timelike Curves." In *General Relativity and Gravitation 1992: Proceedings of the 13th International Conference on General Relativity and Gravitation*, edited by R. J. Glister, C. N. Kozameh, and O. M. Moreschi, 295. Bristol, UK: IOP Publishing, 1993.

Thorne, Kip S., and Frank Tipler. "A Cosmological Dialogue between Caltech Cosmologist Kip Thorne & Tulane Cosmologist Frank Tipler on The Physics of Immortality." *Skeptic Magazine* 3, no. 4 (1996): 64–67.

Tipler, Frank J. "Causality Violation in Asymptotically Flat Space-Times." *Physical Review Letters* 37, no. 14 (October 4, 1976): 879–882.

————. "ETI Beings Do Not Exist." *Quarterly Journal of the Royal Astronomical Society* 21 (1980): 278.

————. *The Physics of Immortality: Modern Cosmology, God and the Resurrection of the Dead.* New York: Doubleday, 1994.

————. "Rotating Cylinders and the Possibility of Global Causality Violation." *Physical Review D* 9, no. 8 (April 15, 1974): 2203–2206.

Travis, John. "Could a Pair of Cosmic Strings Open a Route into the Past?" *Science* 256 (April 1992): 179–180.

Tryon, Edward P. "Is the Universe a Vacuum Fluctuation?" *Nature* 246 (1973): 396–397.

Van Stockum, Willem Jacob. "The Gravitational Field of a Distribution of Particles Rotating about an Axis of Symmetry," *Proceedings of the Royal Society of Edinburgh*, 57 (1937).

————. "A Soldier's Creed (by a Bomber Pilot)." *Horn Book*, Christmas 1944.

Visser, Matt. "From Wormhole to Time Machine: Remarks on Hawking's Chronology Protection Conjecture." *Physical Review D* 47, no. 2 (January 15, 1993): 554–565.

————. *Lorentzian Wormholes: From Einstein to Hawking.* New York: American Institute of Physics, 1995.

————. "Traversable Wormholes: Some Simple Examples." *Physical Review D* 39, no. 10 (May 15, 1989): 3182–3184.

————. "Traversable Wormholes: The Roman Ring." *Physical Review D* 55, no. 8 (April 15, 1997): 5212–5214.

————. "Van Vleck Determinants: Traversable Wormhole Spacetimes." *Physical Review D* 49, no. 8 (April 15, 1994): 3963–3980.

————. "Wormholes, Baby Universes, and Causality." *Physical Review D* 41, no. 4 (February 15, 1990): 1116–1124.

————. "Wormholes, Warpdrives and Other Weirdness." Presentation delivered at the University of Maryland. November 1999.

Visser, Matt, Sayan Kar, and Naresh Dadhich. "Traversable Wormholes with

Arbitrarily Small Energy Condition Violations." *Physical Review Letters*, 90, no. 20 (2003).

Wali, Kameshwar C. "Chandra: A Tribute." In *Black Holes and Relativistic Stars*, edited by Robert M. Wald, 269–272. Chicago: University of Chicago Press, 1998.

Wang, Hao. *Reflections on Kurt Gödel.* Cambridge, MA: MIT Press, 1987.

Wells, H. G. *The Definitive Time Machine: A Critical Edition of H. G. Wells's Scientific Romance.* [Edited] with introduction and notes by Harry M. Geduld. Bloomington: Indiana University Press, 1987.

———. *Experiment in Autobiography.* New York: Macmillan, 1934.

———. *Outline of History; Being a Plain History of Life and Mankind.* New York: MacMillan Company, 1920.

———. *The Time Machine: An Invention. A Critical Text of the 1895 London First Edition, with an Introduction and Appendices.* Edited by Leon Stover. London: McFarland & Company, 1996.

———. *The Wonderful Visit.* New York: Macmillan and Co., 1895.

Wheeler, John Archibald. *Geons, Black Holes, and Quantum Foam: A Life in Physics.* With Kenneth Ford. New York: W. W. Norton & Company, 1998.

Whitehead, Alfred North. *Science and the Modern World.* Cambridge: The University Press, 1929.

Woolf, Harry, ed. *Some Strangeness in the Proportion: A Centennial Symposium to Celebrate the Achievements of Albert Einstein.* Reading, MA: Addison-Wesley, 1980.

"Wormholes and Time Machines." *Science News* 134, no. 19 (November 5, 1988): 302.

Yourgrau, Palle. *The Disappearance of Time: Kurt Gödel and the Idealistic Tradition in Philosophy.* Cambridge: Cambridge University Press, 1991.

Zöllner, Johann Carl Friedrich. *Transcendental Physics: An Account of Experimental Investigations from the Scientific Treatises.* Boston: Colby & Rich, 1881.

INDEX

Page numbers in *italics* refer to diagrams;
page numbers beginning with 311 refer to notes.